General Systems Theory
An Introduction

Lars Skyttner
University of Gävle/Sandviken, Sweden

First published 1996 by
MACMILLAN PRESS LTD
Houndmills, Basingstoke, Hampshire RG21 6XS
and London
Companies and representatives
throughout the world

ISBN 0-333-61833-5

A catalogue record for this book is available
from the British Library

Printed in Great Britain by
Antony Rowe Ltd, Chippenham, Wiltshire

Contents

Preface

As a lecturer in Sweden working within the area of informatics and systems science I have many opportunities to present and recommend international papers and literature. This task is especially stimulating when I act as supervisor for my final-term students as they prepare their degree theses. However, both my students and I are aware that the suggested books are for the most part hidden in some far-off university. The few books that are to be found in the local bookstores and libraries are devoted mainly to the presentation of a single theory and give only a hint of all of the many other theories.

I therefore embarked on the task of presenting a summary of the most prevalent systemic ideas and concepts, well aware of the importance of facilitating access to contemporary theories, especially for those students preparing their theses. In the early phase of this work I also became aware of the lack of easily accessible introductory books within the area. My original intention had then to be revised so as to allow for a presentation of at least the most important subareas of the somewhat dispersed knowledge in informatics and systems science. One of these subareas is information and communication theory. Another is cybernetics, which today is part of General Systems Theory and is defined as 'communication and control in man and machines'. The need for knowledge about basic concepts in information and communication theory is obvious here. This and other main knowledge areas related to systems theory are therefore presented in their own chapters.

Although the selection of theories for this book is inevitably subjective, I hope that my years as a lecturer have given me a reasonable feeling for what should and should not be included. While the following pages are written for students of systems science, they should also appeal to students of related disciplines who want to examine the relevance of this field of knowledge to their own specialism. My hope is that the contents of this book will also offer something of interest for the general reader.

Do we really need system theories? Their claim to be part of a universal science has evoked criticism: a 'theory of everything' has no real content and must of necessity be superficial. However, to be honest, the attempt to gain more abstract and general comprehension must sometimes be made at the expense of concrete and particular exposition. But it is also true that the most general theory, which explains the greatest range of phenomena, is the most powerful and the best; it can always be made specialized in order to deal with simple cases.

The why and how in defence of Systems Theory is presented in the following chapters. When applied sensibly, this theory will make us conscious of the far-reaching interconnections and complexity of our existence. It will show the consequences of adopting solutions that are too spontaneous and too simple and should help us to speak in terms that are understandable in fields as removed from each other as agriculture and astrophysics. Furthermore, it should be recalled that systems theory and its applications emerged out of a need to solve real world problems.

All who attempt to solve problems, make recommendations and predict the future, need theories, models and, as a starting point, concepts, which represent the backbone of the task. Theories introduce order and meaning to observations that may otherwise seem chaotic. Good theories should provide a simplified presentation of complex ideas by establishing connections between hitherto unrelated phenomena. They enhance a growing understanding and help us to guide future research. Those searching for useful ideas among these pages must however realize that the benefit of a certain theory has nothing to do with whether it is 'true' or not – 'truth' is a quality that is undefinable. What we can define is usefulness in relation to our need; different needs obviously demand different theories.

To students asking for a definition of what a good theory is, I recommend the following uncomplicated 'theory'. A good theory is a model that helps you to explain in a simple manner what you are striving to achieve. More scientifically, a theory may be defined as a model concerning our inner or outer environment (or most often a part of it) and some rules which relate entities in the model to observations of reality. The theory can be seen as good if it both satisfies a careful description of a large amount of observations based on a few arbitrary elements and makes reasonable prognoses of future situations. Nearly always, simplicity is the mark of the good theory. But beware – there is no way to prove that a better theory does not exist.

What has not been explicitly written in the chapters of this book, but possibly can be 'read between the lines', is quite naturally a main concern for the author. That this regards a world view where human existence is not guided by a blind faith in computer 'fixes' and 'big' science will, it is hoped, be understood by the observant reader. If the reader feels comfortable with the theories and approach of this book, he or she is running the risk of becoming involved. Any such involvement will further enhance the meaning of this venture.

Finally, I feel compelled to confess two misjudgements when this book was planned: underestimation of the work effort involved and overestimation of my own working capacity. Without these misjudgements this book would never have been written, an experience no doubt shared by many authors. So now the book exists; whether for good or ill can only be judged by the reader.

Lars Skyttner
Helgeåkilen 1995

Part 1: The Theories and Why

1 The Emergence of Holistic Thinking

- **The scholastic paradigm**
- **The Renaissance paradigm**
- **A mechanistic world view and determinism**
- **The hegemony of determinism**
- **The age of relativity and quantum mechanics**
- **The systems age**

'Reality is not only stranger than we conceive but stranger than we *can* conceive.' (*J.B. Haldane*)

While man and his situation are the central focus of all social and humanistic sciences, each science pursues its studies from a certain point of view. Political science concentrates on the society's political and administrative organization. Business economics are concerned with the commercial organization, geography with the physical structure and philosophy with the pattern of thought, views of life and ideologies, to name some examples.

Systems science too has its specific point of view: to understand man and his environment as part of interacting systems. The aim is to study this interaction from multiple perspectives, holistically. Inherent to this approach is a comprehensive historical, contemporary and futuristic outlook.

Systems science, with such an ambition and with its basic Systems Theory, provides a general language with which to tie together various areas in interdisciplinary communication. As such it automatically strives towards a universal science, i.e. to join together the many splintered disciplines with a 'law of laws', applicable to them all and integrating all scientific knowledge. Systems science can promote a culture wherein science, philosophy and religion are no longer separated from each other.

To engage oneself in systems science is therefore a highly cross-scientific occupation. The student will come in contact with the many different academic disciplines: philosophy, sociology, physics, biology, etc. The consequent possibility of all-round education is something particularly needed in our over-specialized society.

Contributions concerning all-round education include thoughts put forward by a number of distinguished people. ***François Voltaire*** once said: 'Education is the only quality which remains after we have forgotten all we have learned'. ***Oscar Wilde*** said in one of his plays: 'Education is a good thing but it ought to be remembered that nothing which is worth knowing can be taught.' A Swedish proverb tells us that: 'Education is not something which can be learned; it is something you acquire.'

In the following pages some Western system-theoretical outlooks and theories will be presented together with central concepts (the Eastern world has its own tradition although science is an offspring of Western civilization as a whole). Some philosophical aspects will also receive attention. The broad spectrum of knowledge will be introduced according to the funnel method: much will be poured in, but the output will be a defined flow of systems knowledge.

The natural starting point should be in the golden age of Greece, the cradle of Western modern human science. Beginning in the Middle Ages will (besides keeping the number of pages down) suffice to provide the background necessary to understand the origin of systems thinking and the subsequent development.

The scholastic paradigm

First we must realize that beliefs and knowledge in any era are influenced by concomitant time-dependent paradigms. That the medieval world view could be described with the help of the **scholastic paradigm** satisfied contemporary needs. Although this paradigm may be characterized as prescientific, it was a complete philosophy which wove together morality and heavenly systems with physical and worldly systems, creating one entity. This amalgamation was based on the following propositions:

- Nature was alive and thus mortal, vulnerable and finite.
- The universe and the nature of time was possible to understand.
- Salvation of the soul was the most important challenge.
- Natural sciences were subordinate to theology.
- The goal of science was to show the correlation between the world and spiritual truth.
- Knowledge was of an encyclopaedic nature, classified and labelled.
- The structure of society was influenced by Heaven and reflected a divine order. The cruciform mediaevial city was not only functional, in addition it was a religous symbol.

Scientific development was thus acknowledged only when it supported religion. The existing method with which to explain the complexities of phenomena was insight or revelation; curiosity as such was a sin. Observation, recording, experimentation and drawing objective conclusions were not encouraged. Nature was viewed as an organism created by God; to destroy Nature was to commit a sin. The natural forces were beyond human control; any protection from them would come from God or from witchcraft. Natural phenomena not understood were given a supernatural explanation. Goal-seeking, *teleology*, was built into nature: stones fell to earth because they belonged to the earth and strove to join their origin.

For the second century AD astronomer ***Ptolemy*** the Universe was of a static nature. No difference was made between reality and dream, between fact and judgement. *Alchemy* was not distinguished from chemistry, nor *astrology* from astronomy. Reason was often regarded as something irrelevant or offensive to the mysterious existence. The connection with reality was unformulated, imprecise, implicit and indeterminate. In physics, for example, one spoke about the five (later a sixth) basic substances. They were:

- Earth
- Air
- Fire
- Water
- Quintessence, including ether
- (Magnetism)

Psychology as a formal science was unknown. Mental qualities, such as *satanic, demonic, human, angelic, divine,* were nevertheless recognized, as were the following manifestations.

Deadly sins:	Cardinal virtues:
	- Justice
- Pride	- Prudence
- Covetousness	- Fortitude
- Lust	- Temperance
- Envy	Divine virtues:
- Gluttony	- Faith
- Anger	- Hope
- Sloth	- Love

(Note that the virtues balance the sins.)

The Greek physician ***Galenos*** (131-201) produced a classification of human beings. According to him, each individual belonged to one of four classes defined by what kind of 'body fluid' was predominant. A certain connection between body fluid and type of personality was considered to be highly significant.

Dominant fluid:	Type of personality:
– Blood	– Sanguine
– Yellow gall	– Choleric
– Black gall	– Melancholic
- Slime	– Phlegmatic

An upset in the balance between the bodily fluids was considered to be the cause of an illness.

Despite of prevailing mysticism, it would be a mistake to consider the mentality of the Middle Ages as primitive. Behind this disregard for the physical world and the world of men lay the image of human existence as a trial. Life was considered to be a journey to heaven. The seemingly austere existence was abundantly compensated for by a rich mental life and a far-reaching spiritual imagination.

The Renaissance paradigm

With the coming of the 16th century the prescientific stage is succeeded by one in which science is acknowledged as capable of describing phenomena, as a route to knowledge. Science itself becomes a source for the development of new technologies. A growing respect for facts tested in valid experiments and a proficiency in the communication of knowledge and opinions emerges. Teleological explanations of observed regularities in human environment (the idea that physical systems are guided by or drawn towards a final goal), earlier seen as a *norm* for various phenomena, are gradually abandoned. In place of those, *laws of Nature* come to be formulated on a mechanical basis. By this means only factors directly influencing the course of events are considered.

A new possibility to cope with human existence is introduced with the emergence of increased knowledge in astronomy. With the discoveries of ***Nicolaus Copernicus*** (1473-1543) the geocentric world view is slowly abandoned in favour of a new heliocentric theory for the movements of celestial bodies. Influenced by earlier aesthetic preferences he continues to consider all planetary movements to be perfectly circular. Thoughts about an infinite Universe and world multiplicity vindicated by the philosopher ***Giordano Bruno*** (1548-1600) are considered to be so provocative by the church that he is sentenced to death and burned at the stake. ***Tycho Brahe*** (1546-1601) develops a newly elaborated technique for observation of planetary movements thereby improving the theory. His achievement is implemented by ***Johannes Kepler*** (1571-1630) to prove the elliptic nature of planet orbiting (The three laws of Kepler). Through the invention of the telescope by ***Galileo Galilei*** (1564-1642) it is possible to have a more realistic perspective on the planet Earth. The Earth can no longer be seen as the centre of all phenomena when it is one among several planets moving around the sun. The discovery of huge numbers of stars proves that the universe is both larger and more diverse than decreed by the Church and theologians. Teleological explanation of motion is discarded and motion is now seen as a force acting on bodies rather than these body's striving to join an origin. In the thoughts of Galilei we see the beginning of the mechanistic world view and the separation between religion and science. 'The world of nature is the field of science.'

Thanks to his experimental and mathematical approach, Galilei is considered to be the first modern scientist. As a researcher he differentiated between *quantitative* and *qualitative* properties. The latter, like colour, taste, and smell were descriptions for things existing only in our consciousness and therefore unfit for use within science (which had to be pursued by universal data originating from the objects).

Another researcher, ***René Descartes*** (1596-1650), contributes his integrated philosophy from chaos to cosmos. He extends the separation between religion and science to one between body and mind, *dualism*. Descartes differs between the body which belongs to the *objective* world of physical reality and that which belongs to the *subjective* world of the mind with its thoughts and feelings.

From here on, the Western religious tradition holding human beings as something unique in this world and perhaps in the universe, begins its implacable retreat. Human consciousness no longer mirrors a divine origin, only itself.

Most of the natural phenomena surrounding man seems however still to be inexplicable, i.e. without apparent causation. The explanations offered were of a purely superstitious nature. In spite of this, it is generally believed, as a principle if not in practice, that a complete understanding of the world is possible. When the Renaissance scientist looks about he sees his own world as a relatively small island of certainty surrounded by a sea of accepted mystery.

The birth of modern science must be seen in relation to the power of the church. The influence of the papal theocracy and the religious world view influenced the course of development. It was very little difference between priests and learned men. The trials of Giordano Bruno and Galileo Galilei showed that science was in danger if it interfered with social questions, that is, the domain and the authority of the Pope. Science has to declare itself independent and neutral, and concepts such as impartiality and objectivity soon became its hallmark, influencing modern civilization much more strongly than religion. The religious imperative of man's supremacy over himself is successively superseded by the scientific imperative of the human right to supremacy over nature.

In our own time, at the end of the 20th century, this classic scientific mentality has lost its significance. The concept of objectivity is however still relevant – if we acknowledge its limitations.

The mechanistic world and determinism

In the beginning of the 18th century, the view that we today call the 'scientific world view' is firmly established in European society, albeit dressed in clothes of its own time. Tradition and speculation are replaced by *rationalism* and *empiricism* with the assumption that natural phenomena can and must be investigated and explained. The inexplicable is now only a matter of 'undiscovered science'. The conception is that reality is determined, exact, formulated, explicit and that it is possible to control the natural forces.

The image of the world changes to that of a machine and the ambition of science is to dominate and conquer Nature. Such an entirely material world could be treated as if it were dead, letting man be the possessor and master of his environment, including all plants and animals, and even permitting the expansion of slavery. This world is also separated from the moral world with which it had been one during the mediaeval era. The spiritual and physical order which were synthesized within the Natural Law (now seen as a mathematical/physical entity) are still influencing the whole universe. All the mysteries of nature can now ultimately be explained in mechanistic terms.

The physical world forms a machine wherein every subfunction could be calculated and events in one part of the universe have consequences for all other parts. In this classic determinism, to every effect there is a cause and to every action there is a reaction. Cause and event initiate a chain of interrelated events. In this eternal continuum annihilation of matter/energy is impossible.

'All things by immortal power
Near or far
Hiddenly
To each other linked are
That thou canst not stir a flower
Without troubling of a star.' (***F. Thompson*** 1897)

Astronomy becames the symbolic area for a materialistic world philosophy: a mechanistic universe of dead bodies passively obeying the order of blind forces. Even the general outlook on man changes and is mainly mechanistic. For many, mechanism has come to be the logical opposite to surperstition.

Men and animals are now in principle nothing more than very elaborate mechanical beings. The human heart becomes a pump obeying pure thermodynamical principles within a hydraulic/mechanical system. This mechanistic era is often called the *Machine Age*, a term rooted both in the world view presented here and in the central role played by machines in the industrial revolution.

The most important name in mathematics/physics of this era is ***Isaac Newton*** (1643-1727). In his *Principia* of 1687 concerning gravitation, Newton presents a working mechanistic universe, independent of spiritual order. In Newtonian mechanics the term initial condition denotes the material status of the world at the beginning of time. Status changes are then specified in the physical laws. Known positions and velocities for planets in our solar systems at one specific moment are thus enough to determine their position and velocities for all future time. Newton's laws therefore automatically had determinism built into them.

Pierre Simon de Laplace (1749-1827), a follower of Newton, is famous for his concept, the 'Laplace demon'. This demon knows the position and speed of every particle in the universe at any moment. Using Newton's laws, it calculates the future of the whole universe.

The idea of the universe as a clockwork mechanism is thus established. On this is founded the doctrine of determinism, implying the orderly flow of cause and effect in a static universe, a universe of being without becoming. Carried to its final extreme, *superdeterminism* is embraced by many of the scientists of the time. According to this world view, not even the initial condition of the universe could have been other than it was; it is determined exactly so by a determinism which determined itself.

The hegemony of determinism

A uniform world view is emerging, expressed in mechanistic terms. It is possible to comprehend the universe, at least fundamentally. This clockwork universe, having been wound up by the Creator, works according to the internal structure and the causal laws of nature. The purpose and meaning, the very existence is put outside of the universe itself. The distinction of a clockwork is just that its meaning is external to the machine and only exists in the mind of its creator. As a clockmaker is to a clock, so is God to Nature.

Clockwork is also presented as a central characteristic of the general *principle of causality*: that every effect is preceded, not followed, by a cause. Just as one cogwheel drives and influences the other in a rational way, a measurable cause always produces a measurable effect in any rational system. Also, identical causes imposed upon identical rational systems, always produce identical effects. Thus one cause/effect relation explains all existence, where the first cause was God.

Under these circumstances, the problem of free-will comes to the fore: free will is claimed to be an illusion. Meaning and freedom of choice lose their purpose in a deterministic universe; they are not necessary to explain natural phenomena and human behaviour. The cause explains the effects completely.

On the basis of this mental world view, *reductionism* becomes the predominant doctrine. Reductionism argues that from scientific theories which explain phenomena on one level, explanations for a higher level can be deduced. Reality and our experience can be reduced to a number of indivisible basic elements. Also qualitative properties are possible to reduce to quantitative ones. Colour can be reduced to a question of wavelength, hatred and love to a question of the composition of internal secretion, etc. Thus reductionism is inherent to all main fields of science, as is illustrated below.

– in physics	: the atom with two qualities, mass and energy
– in biology	: the cell, the living building block
– in psychology	: the archetype instincts
– in linguistics	: the basic elements of sound, the phonemes

Reductionism in turn provides a foundation for the analytical method with its three stages.

- Dissect conceptually/physically.
- Learn the properties/behaviour of the separate parts.
- From the properties of the parts, deduce the properties/behaviour of the whole.

Observations and experiments are the cornerstones of reductionist analytical methodology. Another prerequisite of this method is freedom from environment, that is, environment is considered to be irrelevant. The *scientific laboratory* concept standardizes, and thereby excludes, the environment. In this milieu the effect of different variables – those being observed by the scientist – can be studied in proper order without influence from the environment. Here various hypotheses about nature are tested in order to arrive at approximate answers. Here the ultimate scientific activity is exercised; to *describe*, *control*, *predict*, and *explain* the various phenomena. In this activity the scientist is presupposed to be outside of the experiment. The observer is not involved, at least ideally. The lodestar of the scientist becomes *non-intervention*, *neutrality* and *objectivity*.

The basic metaphysical presumption behind the concept of the laboratory is that nature is neither *unpredictable* nor *secretive* and that it is *computationally reversible*. Predictability implies that the same laws of nature are valid in all parts of the universe. It also implies that the physical states are influenced by laws, but not vice versa. By non-secrecy is meant that all aspects of nature are in principle possible to reveal, albeit that this will sometimes take an extremely long time. The same experiment performed by different observers in different parts of the universe and at different times should always give the same results (*intersubjectivity* and *repetitionality*). Dissimilar results are attributed to human deficiency or deception and will be corrected through better precision of the experimental design. Computational reversibility implies that, given all necessary knowledge, it is possible to calculate what happened in a previous instance, that is, that nothing changes with time.

Through analytical science *The Scientific Method* is established with its own approach following the order presented on the next page.

- reduction of complexity through analysis
- development of hypotheses
- design and replication of experiments
- deduction of results and rejection of hypotheses

This methodology, albeit still with its basic metaphysical assumptions, now becomes the cornerstone of empirical science. It entails a rational, empirical process of inquiry from observation to the formulation of hypotheses and further via experiments to theory. Its strength is its exclusive consideration of relevant fact for what is in focus. An examination of weight thus entirely excludes the colour of the investigated object.

Thus the aim of the method is to bring about a fixed path reasoning appropriate for all kinds of problems. The person who uses it can be assured that he has not been outwitted by nature to believe something that he actually does not know. Note, however, that a scientific accomplishment obtains a value only when it is unrestrictedly and officially communicated to others. Thanks to this implied fifth and imperative step of the methodology, comments and corrections of the result can be fed back to the researcher. This will initiate new ideas and experiments which in turn ensure that the accumulation of knowledge never halts.

Classic empirical science is able to produce not only theories explaining existing phenomena but also theories revealing phenomena not yet discovered. It can even use methods which create unexplained theories in search of phenomena. Abstract elegant theories waiting for a practical application are part of the history of science.

This scientific method lays the ground for a certain kind of mentality and a marked homogenous world view based on the concepts of *empiricism*, *determinism* and *monism*. While empiricism is the doctrine that the universe is best understood through the evidence confronting our senses, determinism is the belief in the orderly flow of cause and effect. Monism implies the inherent inseparability of body and mind, a prerequisite in all European thinking. The above concepts taken together are often referred to as the *Scientific Paradigm*. In the study of electricity, magnetism, light and heat the Scientific Paradigm has great success. Within a short time general mathematical laws are formulated which show the interrelationship between the different areas.

Human optimism grows rapidly: science is expected to give the ultimate answers to questions within all areas. *Scientific positivism* with its demand for 'hard facts' acquired through experience is brought into fashion by ***Auguste Comte*** (1798-1857). Concepts like cause, meaning and goal are weeded out of the natural sciences. Only a reality possible to observe with our senses and possible to treat logically can be accepted as a basis for reliable knowledge. The role of the scientist should be that of the objective observer, explaining and predicting. The collection of absolute facts and the quantification of these are the main occupation of the scientist.

This positivist mentality can be summed up using the following concepts.

- **Philosophical monism** Body and mind are inseparable.
- **Objective reality** A reality possible to experience with our senses.
- **Nominalism** All knowledge is related to concrete objects. Abstractions lack a real existence.
- **Empiricism** All knowledge is founded on experience.
- **Anti-normativism** Normative statements do not belong to science as they are neither true nor false.
- **Methodological monism** Only one method of scientific research exists, that given us by the scientific paradigm.
- **Causal explanations** Goals, intentions and purpose are irrelevant.

At the end of this era of classical determinism, the mechanistic interpretation of thermodynamics leads to new insights. The two main laws of thermodynamics are formulated through works of ***Rudolph Clausius*** (1822-1888), ***William Kelvin*** (1824-1907), ***Ludwig Boltzmann*** (1844-1906) and ***James Maxwell*** (1831-1879), the originator of Maxwell's demon. This is a metaphysical thermodynamic being who apparently neglects the second law by decreasing the entropy into an isolated system. The concept of *entropy* is introduced as an abstract mathematical quantity, the physical reality of which retains a shroud of mystery.

The **first law of thermodynamics** says: The total energy in the universe is constant and can thus be neither annihilated nor created. Energy can only be transformed into other forms. (The principle of conservation of energy with regard to quantity.) In a sense, this law had already been formulated 500 years B.C. by the Greek mathematician Pythagoras who said 'everything changes, nothing is lost'.

The **second law of thermodynamics** states that all energy in the universe degrades irreversibly. Thus, differences between energy forms must decrease over time. (The principle of degradation of energy with regard to quality.) Translated to the area of systems the law tells us that that the entropy of an isolated system always increases. Another consequence is that when two systems are joined together, the entropy of the united system is greater than the sum of the entropies of the individual systems.

Potential energy is organized energy, heat is disorganized energy and entropy therefore results in dissolution and disorder. The sum of all the quantities of heat lost in the course of all the activities that have taken place in the universe equals the total accumulation of entropy. A popular analogy of entropy is that it is not possible to warm oneself on something which is colder than oneself. The process of human ageing and death can serve as a pedagogic example of entropy. Another common experience is that disorder will tend to increase if things are left to themselves (the bachelor's housekeeping!).

Inasmuch as there is a mathematical relation between probability and disorder (disorder is a more probable state than order), it is possible to speak of an evolution toward entropy. Below some well-known expressions illustrates this process.

Probability	Improbability
- Disorder	- Order
- Disorganized energy (heat)	- Organized energy
- Heat (low–grade energy)	- Electricity (high-quality energy)
- Entropy	- Syntropy

The above process derives from the second law of thermodynamics and has had a tremendous impact on our view of the universe. One consequence is to experience the world as *indeterministic* or as chaotic. The ultimate reality is the blind movements of atoms whereby life is created as a product of chance, and evolution is the result of random mutations. Another is that the Newtonian world machine has a persistent tendency to run down; the Creator must wind up the celestial clockwork from time to time. Any event that is not prohibited by the laws of physics should therefore happen over and over again.

Today we can see how these perspectives, together with the image of the inevitable death of the universe, have significantly influenced philosophy, art, ethics, and our total world view. This image has inflicted upon the Western culture some form of paralysis. For the generations of researchers nurtured *via* this period's mentality, a physical eternity without purpose seemed to be the basis for all reality. For these people the Universe could be described as 'big and old - dark and cold', quoting the contemporary geologist George Barrow. The French physician ***Léon Brillouin*** (1889-1969) sums everything up in his question 'How is it possible to understand life when the entire world is organized according to the second law of thermodynamics which points to decay and annihilation?'

The era of determinism coincides with both the era of machines in the industrial revolution and the conservative Victorian culture. Human skills are increasingly taken over by machines; the remaining manual tasks are broken down into a series of simple and monotonous manipulations. This dehumanization of productive effort and the subsequent alienation of the worker gives rise to mental phenomena such as Marxism-Leninism.

The deterministic era can also be named the age of *scientism*, with reference to the belief that only concepts which can be expressed in the language of the exact natural sciences and proven by quantification have a reality. It assumes the existence of an objective reality, including dichotomies contrasting man and nature, mind and matter, facts and values. Its primary concern is to discover truth, regarding questions of values and needs as outside the realm of scientific inquiry. Scientism is also synonymous with the 'objective' mode of presentation of results, used by many researchers of this era. That courage, despair and joy are important prerequisites for a successful result is neglected - for entirely subjective reasons.

In the deterministic interpretation of the second law of thermodynamics it is possible to find the roots of the pessimism prevailing at the turn of the century. The sun is exhausting its life-giving resources, the earth is approaching a new glacial period and the society is declining. Inferior army discipline, general decadence, falling birth rate, spread of tuberculosis are all visible effects of increased entropy. Emotionally, cosmic and physical values are never separated from a human system of evaluation. The resulting gloominess, the *fin de siècle* mode, is excellently presented in European literature of this period.

While a 300-year-old attitude towards reality draws to its end, the dissolution of determinism gives room for new impulses and new perspectives.

The age of relativity and quantum mechanics

The first fatal blow to determinism with its static view of the universe comes from ***Albert Einstein*** (1879-1955) in 1905, in his *special* theory of relativity. An event is defined with four numbers: three for the position in space and one for time. These constituents do not exist individually; it is not possible to imagine time without space, or vice versa. When a star is observed at a distance of one hundred light-years, the star is not only this far away in space but it is also observed as it was one hundred years ago. The four-dimensional space with its space/time continuum is introduced.

The contradiction between this theory and Newton's theory of gravitation poses a problem. Einstein solves it in 1915 by introducing the *general* relativity theory, where gravitation is a consequence of the nonflat curving space/time caused by the content of mass and energy. The mass of the sun curves the space/time into a circular orbit in the three-dimensional world even if it is a straight line in the four-dimensional world. Einstein's synthesis of the fundamental quantities of time, space, mass, and energy is confirmed first in the 1930s through astronomical observations.

For the general public living in the first part of the twentieth century, the scientific world view represented by Einstein's theories was sometimes more than incomprehensible. A contemporary view of the general relativity theory may be found in the following limerick:

There was a young lady girl named Bright,
Whose speed was far faster than light,
She travelled one day,
In a relative way,
And returned on the previous night. (***R. Buller***)

Another death blow to determinism was *quantum theory*. It had been enunciated already 1901 by the German physicist ***Max Planck*** (1858-1947). With this theory the classic concept of mechanics starts its reformulation. In 1927 it is ***Werner Heisenberg*** (1901-76) who frames *the uncertainty principle*: it is fundamentally impossible to simultaneously define position and velocity for a particle. Heisenberg's principle must be considered a special case of the *complementarity principle*, also articulated in 1927 by ***Niels Bohr*** (1885-1962). This states that an experiment on one aspect of a system (of atomic dimensions) destroys the possibility of learning about a complementarity aspect of the same system. Together these principles have shocking consequences for the comprehension of entropy and determinism.

The new mechanics, *quantum mechanics*, thus includes indeterminism as a fundamental principle when it focuses on the atom and its particles. In this small-scale system, the predominant and special circumstances are explained with the help of quantum theory. This theory concerns *probabilities* rather than certainties. Although concerned solely with extremely small particles, the theory reveals some extraordinary circumstances in physics. One is 'A spooky action at a distance', as Einstein called the *spectral effect*. A pair of correlated particles which have at one time been connected continue to influence each other instantly even after they have moved to separate parts of the universe.

Another circumstance is that electrons will not jump from one energy level to another while they are being watched, the *zeno effect*. This illustrates a basic phenomenon within quantum physics; the interpreter and the interpreted do not exist independently. Thus, interpretation is existence and existence is interpretation.

While quantum theory is not the final answer in physics, it has definitely opened a completely new way of thinking; its impact on the perception of reality and our world view should not be underestimated. Today, most scientists agree on a world view in which global determinism points in a main direction; they agree that local development determines its own non-predictive path, open to causal influences coming from both lower and higher levels.

The predominant cosmological view, called the *standard model,* tells us that the universe is expanding and has as its starting point in time the *big bang* of 15 billion years ago (the greatest effect of all with no cause!). The universe has then developed from an incredibly tightly-packed system, a *singularity*, where the natural laws as we know them did not exist. This condition cannot be described with the help of *either* a theory of relativity *or* the quantum theory. These can at most be seen as components of a not yet existing final theory.

We are now nearing the end of the 20th century. What was begun by Galileo, continued by Newton and finished by Einstein has over time inspired even poets:

> 'Nature and nature's laws lay hidden in night
> Let Newton be God said and all was light'. (***Alexander Pope***)
>
> 'But then the devil cried that Einstein had to do
> his work and reestablish status quo'. (***John Collings***)

These small poems implicitly question whether we can understand the world surrounding us and theories about it. Theories such as the quantum theory cannot actually be proved. If they are mathematically consistent and observations coincide with predictions, the probability is however high that they describe reality reasonably well. Today, the rules of quantum theory have been around for a long time and must be considered neither wrong or incomplete. But modern science based on quantum theory has come to realize that it is impossible to conclusively describe and understand the natural world. To this may be added that even if modern science was able to explain *how* the Universe is structured, it cannot say *why*.

Scientists today tend to agree that when we formulate the theories of the atomic world, we are doing it *vis-à-vis* not the reality but rather our knowledge regarding reality. Physics, for example, does not claim anything about something actually existing, but rather informs our knowledge concerning that which we claim exists. The models of physics no longer explain, they only describe. Therefore, in a way, fundamental physics today is a matter a of philosophy, while cosmology has been a kind of scientific poetry.

A consequence of this attitude is that it is possible to claim that the world only exists in the spectator's mind, that an observation is dependent upon the observer. This philosophical shattering of reality echoes the claim of ***Immanuel Kant*** (1724-1804) that the concepts of space and time were necessary forms of human experience, rather than characteristics of the universe. Kant considered that it is not only the consciousness which adapts to things, but things also adapt to the consciousness. This kind of physical idealism is well expressed in another limerick:

> There once was a man who said 'God
> Must think it exceedingly odd
> If he finds that this tree
> Continues to be
> When there's nobody else in the quad.' (***Ronald Knox***)

The view that only one truth about reality exists and that the various scientific disciplines describe different parts of it is no longer tenable. What exists is only subjective and often contradictory conceptions of reality. The decline of the illusions of the pre-Einstein natural science shows that not even scientific results are absolute. In due time they are replaced by theories and models having an extended descriptive and predictive value. Present-day knowledge is only the best description of reality we have at the current moment in time.

Werner Heisenberg is reported to have said: 'A quantum world does not exist. The only thing which exists is our abstract description of the physical reality.' ***Niels Bohr*** (1885-1962) also said: 'Physics is only about what we can say concerning nature.' There is no point in asking how matter could be constituted behind our observations of it, as these are the only evidence we can ever have. According to this view, quantum theory should not be understood as a description of the world, but rather as an instrument enabling the human mind to make predictions and calculations.

Albert Einstein took a slightly different view when he said: 'The firm laws of logic are always valid, and nature's laws are indifferent to our attitude.' Thus Einstein claimed that the world exists independent of human beings and that it is only in part comprehensible. Einstein's pursuit of the old rationalist tradition in Western science that reality has an objective existence independent of the observer is thus today questioned by many researchers.

The multiple perspectives, issuing from the modern, relativistic science, have actualized the dualism between substance and awareness, the classic body/mind problem. Our conventional definition of self-consciousness includes totality and consistency in time and space. Such self-consciousness can be achieved only by a creative human intelligence. Quantum physics claims that consciousness *per se* may be seen as the particle's *mental* existence in wave form defined by cooperation, interference and overlapping. It exists everywhere and has knowledge of what happens in other places. The particle's *physical* existence is its permanence as matter with mass and position in space. On the basis of the above we can identify the following internal respective external opposites:

- consciousness	- body
- subject	- object
- individual	- environment
- culture	- nature

A number of proposals taken from the area of relativity theory and quantum physics and are presented below. Many of these are paradoxical.

- There is an infinite number of worlds and we exist parallel in them.

- Time goes both backward and forward at the same time.

- Matter and consciousness are the same thing.

- A particle exists in several places at the same time when manifested as a wave form. Although it can only be observed in one place at a time, it does exist in several spaces simultaneously.

- Quantum physics concerns probabilities. Quantum wave functions express all probabilities simultaneously. When someone observes, the probability becomes a reality with fixed properties. Other possibilities vanish.

- In the world of quantum physics everything is interconnected. Everything exists everywhere simultaneously, but can only be observed as an object in one universe at a time.

- The quantum wave is a connection through all time both in future and past time.

- What we remember of past times has been determined by something in the future. Both past and future have existed before, the future in a parallel universe.

- When we choose to observe something, we create and influence it.

- Observations create consciousness and consciousness creates the material universe.

- The existence of matter and consciousness is the same thing.

- The existential basis for all matter is meaning.

- Radio-transmitted music confined in the form given by the radio wave exists as a potentiality; it is heard only when the receiver is turned on.

- Quantum fields of potential information are everywhere omnipresent. Their meaning is existence. To change the meaning changes the existence.

- The mental and the physical world are two sides of the same coin. They are separated by consciousness only, not by reality.

- Meaning and purpose are inherent parts of reality, not an abstract quality in the human mind.

Quantum theory has seriously undermined science's faith in an external, material reality and has implied a repudiation of scientism and a rigorous positivistic empirical science. The potential of dead matter to produce living matter and consciousness signifies a recognition of purpose, of creation and self-organization. The function of living matter is apparently to expand the organization of the universe. Here locally decreased entropy as a result of biological order in existing life is invalidating the effects of the second law of thermodynamics, although at the expense of increased entropy in the whole system.

Strict determinism is no longer valid; the development of our universe is decided both by chance and necessity, by random and deterministic causes working together in entropy, evolution, continuity and change. These universal principles – sometimes called **syntropic** (***Fuller*** 1992) – counteracting decay and destruction (the second law of thermodynamics), will create a new and more flexible world view.

Another challenge to positivistic science is the idea that the universe itself is a living phenomenon, irrespective of its organic inhabitants. The creation of new stars, their growth, reproduction and death, together with their metabolism, justify the use of the term 'living' in the eyes of many scientists.

The concept of value is not inherent to science; classical science never asks why or what for. Nor is speculation as to the cause, meaning and ultimate goal an attribute of its method. The second law of thermodynamics, expressing the diffusion and deterioration of matter and existence, has long represented the classical science mentality and influenced the construction of methods and instruments. Today, with a growing awareness of the universe undergoing a creative and problem-solving evolution, values can add new and fruitful dimensions to classical science.

On the basis of the above outline of this scientific development and the consequences thereof for the present-day world view, some observations can be emphasized. A gratifying, and astonishing, fact is that the classic natural laws formulated by Newton, for example, are still going strong. While piece after piece has been added to the theoretical building by new generations of scientists, it has not yet been necessary to demolish its main structure and start from scratch again.

The Newtonian gravitational theory has influenced Einstein's theory of relativity. Through Einstein's theories Newton's equations have become more complex; Newton's original theory is nonetheless still valid and gives us in most cases very good approximations. Newton's mechanics has now become a 'special case' within Einstein's theory of relativity. The counter-intuitive sub-atomic paradoxes of quantum physics do not interfere with the common sense of everyday life, although they are very extensive in for example microelectronics.

Regarding the relation between relativity theory and quantum theory, the latter suggests that space and time are approximate concepts, which may have to be abandoned when the infinitely small is contemplated. Thus large-scale mechanics and quantum mechanics have been forced to co-exist, because neither is any good at explaining the other.

Another observation is that the classical division of the disciplines was to a great extent conditioned by – but also reflected – the order of nature, mind and society of its time (that is, the well-organized Victorian society). This is expressed by Comte's hierarchy of development in science with its three stages.

- The theological stage (corresponding to the scholasticism)
- The metaphysical stage (corresponding to the the Renaissance)
- The positive stage (corresponding to the mechanistic era)

At the same time it is possible to see a reductionist hierarchy in the various scientific disciplines when arranged in order according to 'size'.

- Astronomy
- Sociology
- Psychology
- Biology
- Chemistry
- Physics

Further, the various disciplines in science have undergone a similar development and show a parallelism in their development of methods. Every field of human knowledge thus passes through distinct stages.

- Intuition
- Fact-finding
- Analysis
- Synthesis

Synthesis is a prerequisite for the **systems thinking** of our own time, just as analysis was for the mechanistic era. A system inasmuch as it is a whole, will lose its synergetic properties if it is decomposed; it cannot be understood through analysis. Understanding must therefore progress from the whole to its parts – a **synthesis**. Synthesis takes the steps of analytical science (see p. 10) in reverse order.

- Identify the system of which the unit in focus is a part.
- Explain the properties or behaviour of the system.
- Finally, explain the properties or behaviour of the unit in focus as a part or function of the system.

Synthesis does not create detailed knowledge of a system's structure. Instead, it creates knowledge of its function (in contrast to analysis). Therefore, synthesis must be considered as **explaining** while the scientific method must be considered as **describing**.

Systems thinking **expands** the focus of the observer, whereas analytical thinking **reduces** it. In other words, analysis looks into things, synthesis looks out of them. This attitude of systems thinking is often called **expansionism**, an alternative to classic reductionism. Whereas analytical thinking concentrates on **static** and **structural** properties, systems thinking concentrates on the function and behaviour of whole systems. Analysis gives description and knowledge; systems thinking gives explanation and understanding. With its emphasis on variation and multiplicity, rather than statistically ensured regularities, systems thinking belongs to the holistic tradition of ideas.

Systems thinking is a response to the failure of mechanistic thinking in the attempt to explain social and biological phenomena. As an attempt to solve the crisis of classical science it has formulated new approaches in scientific investigation. Primarily, it dates back to the 1920s when emergent properties in living organisms were generally recognized. Born in biology, it is easy to understand that the systems movement has acquired the major part of its terminology from that area when considering terms like autonomy, survival, etc.

It is now possible to note how the specific tools in the various areas have emphasized the different stages. The tools for the analysis were *par excellence* the microscope and the telescope, tools which must be considered to be reductionist promotive. The tools of the emerging systems age are designed to enhance synthesis and have often taken over the function of the classical laboratory. The computer has become a viable substrate for experimentation. Research in many fields such as nuclear, aerodynamics, biology, chemistry, etc., is now being simulated instead of actually performed. The particle accelerator combines analytic and synthetic properties in a kind of super microscope capable of the resolution of objects less than the diameter of the atomic nucleus. Satellites orbiting the earth give outstanding possibilities for the understanding of global phenomena and for the first time in history humanity has now the opportunity to look upon itself from the outside. Tools with the above-mentioned properties are often called **macroscopes**.

The systems age

In the 1950s, with the introduction of computers, hydrogen bombs and space exploration, large-scale problems began to penetrate Western society. The traffic-system breakdowns, environmental disasters and the nuclear threat were immediately high on the agenda. Society was faced with *messes*, interacting problems varying from technical and organizational to social and political.

It was suddenly realized that many solutions were inadequate when applied to problems which no longer existed in their original form. Change itself, with its accelerating rate, was a major concern. Two hundred years of success for classical science and technology had created a form of development the long-term effects of which apparently were programmed to be devastating for humanity. ***Gerald Weinberg*** states in one of his books that 'science and engineering have been unable to keep pace with the second order effects produced by their first order victories'.

The following examples address some of the problems:

- environmental destruction and climatological changes
- deforestation and desertification
- garbage accumulation, nuclear radiation, water, soil, and air pollution
- acidification, decreasing subsoil water and shrinking ozone layer
- decreasing biodiversity and extinction of species
- population explosion and criminalization
- urbanization, unemployment, and proletarianization
- energy wastage and resource depletion
- motorization and noise pollution
- data pollution, lack of information and knowledge
- commercialization and cultural impoverishment
- mental corruption, drug abuse and AIDS
- environmental ugliness with growing amounts of concrete and asphalt
- bureaucratization, passivisation and dulling of the human intellect

Classical science, with its overspecialization and compartmentalization, had already proved its inability to handle problems of such tremendously increased complexity. Its attempt to reduce complexities to their constituents and build an understanding of the wholeness through knowledge of its parts is no longer valid. Not understanding that the wholes are more than the sum of their parts, scientists had assembled knowledge into islands, extending into an archipelago of disconnected data.

Not long ago physics was regarded an archetype for all genuine science. A reductionary chain was envisaged where psychology was deducted to neurophysiology, neurophysiology to biochemistry and biochemistry in turn to quantum mechanics. Today, modern biology has shown that this kind of reductionism is out of the question. Physics, chemistry and biology have united with each other into molecular biology - a new overarching description system separated from the area of both physics and chemistry.

Many scientists have now realized that the way they had embraced the world was not far-reaching enough to understand and explain what they observed and encountered. As ***Gary Zuchov*** (1979) says in his book *The Dancing Wu-Li Masters*: 'Their noses had been too deeply buried in the bark of a special tree, to be able to discuss forests in a meaningful way.' It was thus accepted that systems are wholes which cannot be understood through analysis inasmuch as their primary properties derive from the interactions of their parts. Furthermore, interaction between systemic variables are so integrated that cause and effects cannot be separated - a single variable could thus be both cause and effect.

Thus awareness grew that everything in the universe - including themselves

– which seems to exist independently was in fact part of an all-embracing organic pattern. No single part of this pattern was ever really separated from another. It was possible to catch a glimpse of a universality of systemic order and behaviour which characterized both living and nonliving systems. That humans now had got access to some of the main design principles of the universe implied that they too were included in the drawings for some very significant ultimate purpose.

Earlier, the alternative to systemic intervention was to suffer the consequences, to endure whatever happened; scientists had too often waited for systems failures to see what these could reveal about the mechanism. Today function, not anatomy, is the main point. The important task is to solve problems in real life. To describe and understand were not values in themselves; their purpose was to enhance the capability for large-scale system prediction and control.

The technicians strove to have things work well, the social scientist to have things behave well. Science was to become more **ethical,** less **philosophical.** To do things, was considered to be more important than to think about them. In these circumstances emerged the new **interdisciplinary** and **holistic** approach. Here holism was an attempt to bring together fragmentary research findings in a comprehensive view on man, nature, and society. In practice it was a search of an outlook to *see better*, a network to *understand better* and a platform to *act better*.

Without hesitation this had it roots in the war-time efforts and the special mentality of **operational research.** This 'emergency-discipline' handled military strategic decisions, resource allocations, optimal scheduling and risk analysis, etc., in a truly pragmatic way, all in order to win the war. Its main guidelines were the following:

- It is not necessary to understand everything, rather to have it under control. Ask what happens instead of why.

- Do not collect more information than is necessary for the job. Concentrate on the main consequences of the task, the small details may rest in peace.

- Solve the problems of today and be aware that prerequisites and solutions soon become obsolete.

Operational research gave rise to the first successful methodology where the problem complex not was disassembled into disciplinary parts and could be treated as one entity by different researchers.

In 1954 the International Society for General Systems Theory, ISGST, was founded. This society later become the International Society for Systems Science, ISSS. Two of the most prominent founders were ***Ludwig von Bertalanffy*** and ***Kenneth Boulding***. Although Bertalanffy had formulated his ideas already in the

1930s, he was not recognized until one of his now classic papers on systems theory appeared in the American journal *Science* in 1950. Then, the idea that systems had general characteristics independent of the scientific areas to which they belonged was both new and revolutionary. Boulding in turn published his well-known system hierarchy in 1956.

The founding team of interdisciplinary scientists, had a shared interest in a universal science. They wanted to link together the many splintered disciplines with a **law of laws** applicable to them all. The following aims were stated:

- to integrate similarities and relations within science;
- to promote communication across disciplinal boundaries;
- to establish a theoretical basis for general scientific education.

Integration should be promoted by the discovery of analogies and isomorphisms and the new science should be a tool with which to handle complex systems. **Analogies** are explanations done by relating something not yet understood to something understood. **Isomorphism** exists when common characteristics, structures, formulas and form of organization are in accordance in different systems. That is, when formally identical laws governing the functioning of materially different phenomena exist. A partial accordance is generally referred to as **homomorphism**. The use of isomorphism made possible the indirect study of systems in terms of other systems (simulation) and the use of content-independent methods within different scientific areas.

Step by step a theory was established: the **General Systems Theory** or **GST**. As a basic science it deals on an abstract level with general properties of systems, regardless of physical form or domain of application, supported by its own metaphysics in **Systems Philosophy**. General Systems Theory was founded on the assumption that all kinds of systems (concrete, conceptual, abstract, natural or man-made) had characteristics in common regardless of their internal nature. These systems could serve to describe nature and our existence.

Expressed in more precise terms, the goal of General Systems Theory can be specified as follows:

- To formulate generalized systems theories including theories of systems dynamics, goal-oriented behaviour, historical development, hierarchic structure, and control processes

- To work out a methodological way of describing the functioning and behaviour of systems objects

- To elaborate generalized models of systems

As an applied science, GST became **Systems Science**, a *metadiscipline* with a content capable of being transferred from discipline to discipline. Its equivalent to the classical laboratory become the computer. Instead of designing experiments with real materials, the computer itself became a viable substrate for experimentation. The use of computers as instruments for calculations, simulations and the creation of a non-existing reality thus brought about a new phenomenon that is neither actual nor imaginary, a phenemenon or mode that was called *virtual*. The computer is a virtual reflection of a non-existing mechanical adding machine. To be precise, it is an abstract entity or process that has got physical expression. In itself, it is a simulation, a simulation which is not necessarily a simulation of anything actual. 'Virtual' is thus a mode of simulated existence, resulting from computation.

The aim of systems science was, however, not to replace, but to complement traditional science. The systems perspective naturally acquired greater significance with the growing complexityof all systems, including and embracing man. Gerald Weinberg (1975) says about systems science that it has '...taken up the task of helping scientists to unravel complexity, technologists to master it, and others to learn to live with it.'

The aim of systems science was however not to replace but to complement the traditional science. The systems perspective quite naturally acquired greater significance with the growing complexity of all the systems including and embracing man. **General systems thinking** based on systems theory became its hallmark with the aim of fostering generalists qualified to manage today's problem better than the specialists. Specific individual methods were developed, many of which included modelling, simulation and gaming.

One of these methods, the **Systems Approach**, in reality an application of Systems Theory, operates in an integrated framework of modern organizational knowledge and management science. The Systems Approach is based on the fundamental principle that all aspects of a human problem should be treated together in a rational manner. It is an attempt to combine *theory*, *empiricism* and *pragmatics* and looks at a system from the top down rather than from the bottom up.

Another method, **Systems Analysis**, adopting a strictly systemic outlook on complex organizations, entered the scientific scene to ensure that no important factors in the structure were excluded. Problems of identifying, reconstructing, optimizing, and controlling an organization, while taking into account multiple objectives, constraints and resources are worked up. Possible courses of action, together with their risks, costs and benefits are presented. Systems analysis can thus be considered an interdisciplinary framework of the common problem-view.

An extension of this method, called **Anasynthesis**, was introduced with the implicit assumption that the more views one can apply to it, the better a problem can be understood. When using this method, *modelling, simulation, gaming, analysis* and *synthesis* are all applied to the development of a system. The method is used iteratively at both the macro and micro levels of large-scale systems. Normally, the outcome is more organized, structured and responsive to real-life requirements than are outcomes of other methods.

Then there is **System Engineering**, a method by which the orderly evolution of man-made systems can be achieved. Hereby the the four Ms - **m**oney, **m**achines, **m**aterials and **m**en - are used in making complex systems in their totality.

A much-discussed method of a more theoretical kind is **System Dynamics**. Developed by ***Jay Forrester*** (1969) it uses dynamic computer models which change in a network of coupled variables. It has been employed to prognosticate the growth of the modern city (Urban dynamics), the development of Western industry (Industrial dynamics), and the global resource depletion (World dynamics).

Closely connected to the above-presented methods, and including them all, is the conviction that man is more the creator of reality than its discoverer. The future has become too complex to foretell or to be planned; it has to be created. Embracing such a pragmatic view on reality, **design** or **redesign** becomes the key concept of the systems perspective when it is about to change the world for the better by building new or improved systems. Design is concerned with how things ought to be, with combining resources to attain goals. This involves processes necessary to *understand the problem*, to *generate solutions* and to *test solutions* for feasibility. Here design is a creative process, questioning the assumptions upon which earlier structures have been built and demanding a completely new outlook. **Systems design** is a formal procedure where human resources, artefacts, techniques, information and work procedures are integrated into a system in order to facilitate its performance. Systems design is the opposite of *systems improvement*, the policy of recovering old systems (***J. van Gigch*** 1978).

A more recent perspective when investigating systems is that of **teleology**, the doctrine that behaviour and structure are determined by the purpose they fulfil. It indicates that systems are guided not only by mechanical forces but also move toward certain goals of self-realization. Here organizations and organisms have their own purposes, while artefacts, e.g. machines, serve the purpose of others but have no such purpose of their own.

Complex systems can thus be studied from many points of view which are seen as complementary rather than competitive. The choice of theoretical approach depends mainly on the type of insight which is sought. A common quality of the named methods is the generation of knowledge necessary for the solving of the problem. The characteristic tools of the domain - computers, telecommunication networks, databases, etc. - are to be found in **informatics**.

One effect of the new approach was that subsets of traditional scientific areas amalgamated, forming new disciplines. A fresh example is the **science of complexity**, where biological organization, computer mathematics, physics, parallel network computing, nonlinear system dynamics, chaos theory, neural networks and connectionism were brought together. This stimulated the definition of new reciprocal systemic qualities: complexity/simplicity and simulative/non-simulative. A new quantification of complexity was also introduced: the complexity of something should be defined as the length of the shortest possible description (algorithm) of this something.

Complexity theory tries to describe how complicated rules sometimes produce simple and organized behaviour, e.g. the ability of living systems to become ever more organized. Its working methodology is non-reductionist: a system is viewed as a network of interacting parts, nearly all the fine details of which are ignored. Regularities and common patterns valid across many different systems are carefully examined. Of specific interest are those conditions which ensure the emergence of evolutionary, self-organizing and self-complicating behaviour. Complexity theory operates somewhere in the zone between the two extremes of complete order and complete chaos.

Also, disciplines more directly related to systems science, such as cybernetics, bionics and C^3I, merit presentation. They make possible a broader perspective concerning the basic underlying principles of structure and behaviour in systems.

Cybernetics was defined in 1948 in a book by ***Norbert Wiener***: *Cybernetics or Control and Communication in the Animal and the Machine.* In cybernetics living systems are studied through analogy with physical systems.

Bionics, the study of living systems in order to identify concepts applicable to the design of artificial systems, was introduced by ***Major Steele*** in 1958. The amalgamation of biology and technique is recognizable in the term. Bionics realizes physical systems through analogy with living systems. Cybernetics and bionics are often said to be the two sides of the same coin.

The acronym $\mathbf{C^3I}$ stands for command, control, communication and intelligence. During the past ten years interest in the operations of social, military and business organizations has grown. Modern managerial systems are based on an interchange between people, organizational entities and technical support. The decision-making situation has often such an innate complexity that in the initial phase it is not possible to define what kind of information is important; the decider usually demands more information than will be useful.

In the extended acronym $\mathbf{C^4I^2}$ the extra **C** stands for computer and the extra **I** for integration, emphasizing the close interconnection between man and computer. Here it is impossible to separate social from technical factors and the human being is always a part of the problem as well as a part of the solution. The adaptation man/machine is a key issue and the system has to be designed around man, his potential and his needs. In spite of access to high-tech decision support, a main point must be the training of human ability to handle the unexpected. Reality always tends to deliver a situation never met before.

Systems science applied as a problem solver in business organizations is sometimes called **management cybernetics**. As such it is often occupied with design of an appropriate organisational structure which includes:

- Specification of the organization's subtasks and partition of work

- Design of communication between the subsystems

- Definition of areas of decision-making and authority

- Design and development of control systems and coordination of efforts toward the organizational goal

Let us recapitulate the emergence of the systems movement with some often-cited words of Kenneth Boulding from 1956:

> 'General Systems Theory is the skeleton of science in the sense that it aims to provide a framework or structure of systems on which to hang the flesh and blood of particular disciplines and particular subject matters in an orderly and coherent corpus of knowledge.'

To that must be added that one of the most important contributions of the systems area is that it provides a single vocabulary and a unified set of concepts applicable to practically all areas of science.

Review questions and problems

1. Learned men of the scholastic era shared the belief that the nature of universe and time was possible to understand. In what field of knowledge could this conviction be studied?

2. What is the main difference between a teleological norm and a law of nature?

3. Why did science as a human activity have to declare itself independent, neutral, and objective from its earliest days?

4. In the deterministic era, the question of free-will was considered irrelevant. Why?

5. What are the most significant metaphysical presumptions behind the concept of the laboratory?

6. The scientific method is associated with five methodological steps. Describe the last step and explain its importance.

7. The old Greek mathematician Pythagoras once said: 'Omnia mutantum, nihil interit' (Everything changes, nothing is lost). How can this be associated with the laws of thermodynamics?

8. Quantum theory has been used in an attempt to explain the classic body/mind problem. How does the train of thought run?

9. Gerald Weinberg states that 'science and engineering have been unable to keep pace with the second order effects produced by their first order victories'. Give some examples of particularly devastating secondary effects influencing modern society.

10. Has systems theory been successful in formulating a law of laws applicable to all scientific disciplines? If so, how does one of these laws read?

2 Basic Ideas of General Systems Theory

- **GST and concepts defining systems properties**
- **Cybernetics and concepts defining systems processes**
- **General scientific and systemic concepts**
- **Widely-known laws, principles, theorems and hypotheses**
- **Some generic facts of systems behaviour**

'On questions of the ends to which means should be directed, science has nothing to say.' (*N. Campbell* 1953)

Each body of theory has its implied assumptions or axioms which in reality are impossible to prove and hence must be accepted as value judgements. The underlying assumptions and premises of systems theory can be traced backward in history. The Greek philosopher, ***Aristotle*** (384-322 B.C.), presented a metaphysical vision of hierarchic order in nature – in his biological systematics. His finalistic, or teleological, natural philosophy represents a rather advanced systems thinking for the time.

Closer to our own era, ***Fredrich Hegel*** (1770-1831) formulated the following statements concerning the nature of systems.

- The whole is more than the sum of the parts.
- The whole defines the nature of the parts.
- The parts cannot be understood by studying the whole.
- The parts are dynamically interrelated or interdependent.

The concept of holism received its first modern appraisal through 'structuralism', a scientific school of thought established by the Swiss linguist ***Ferdinand de Saussure*** (1857-1913). Structuralists studied 'wholes' that could not be reduced to parts. Society was not regarded as a conscious creation; it was considered to be a series of self-organizing structures overlapping each other, with a certain conformity to law. This wholeness regulated the personal and collective will.

After World War I the limits of reductionism were known and the concept of holism already established (particularly in biology). A comprehensive exposition of holism was presented by the Boer general ***Jan Smuts*** (1850-1950) in his book *Holism and Evolution* from 1926. By this book Smuts must be considered one of the most influent forerunners of the systems movement.

In General Systems Theory one of the basic assumptions embraces the concept of order – an expression of man's general need for imaging his world as an ordered cosmos within an unordered chaos. A consequence implicit in this order is the presumed existence of a law of laws which in turn inspired the name of the theory. The systematic search for this law is a main task for General Systems Theory. Another fundamental assertion is that traditional science is unable to solve many real-world problems because its approach is too often narrow and inclined toward the abstract. Systems science in contrast is concerned with the concrete embodiment of the order and laws which are uncovered.

Kenneth Boulding (1964) has formulated five postulates which must be regarded as the starting point for the development of the modern General Systems Theory. They may be summarized as follows.

- Order, regularity and non-randomness are preferable to lack of order or to irregularity (chaos) and to randomness.

- Orderliness in the empirical world makes the world good, interesting and attractive to the systems theorist.

- There is order in the orderliness of the external or empirical world (order to the second degree) – a law about laws.

- To establish order, quantification and mathematization are highly valuable aids.

- The search for order and law necessarily involves the quest for those realities that embody these abstract laws and order – their *empirical referents*.

Other well-known basic assumptions regarding general systems theory as a philosophy of world and life have been summarized by ***Downing Bowler*** (1981). A selection is given below.

- The Universe is a hierarchy of systems; that is, simple systems are synthesized into more complex systems from subatomic particles to civilizations.

- All systems, or forms of organization, have some characteristics in common, and it is assumed that statements concerning these characteristics are universally applicable generalizations.

- All levels of systems have novel characteristics that apply universally upward in the hierarchy to more complex levels but not downward to simpler levels.

- It is possible to identify relational universals that are applicable to all systems at all levels of existence.

- Every system has a set of boundaries that indicates some degree of differentiation between what is included and excluded in the system.

- Everything that exists, whether formal, existential, or psychological, is an organized system of energy, matter and information.

- The Universe consists of processes synthesizing systems of systems and disintegrating systems of systems. It will continue in its present form as long as one set of processes does not eliminate the other.

A short summary of Bowler's assumptions could be expressed in the statement that the design of the macrocosm reflects the structure of the microcosm.

A further perspective on systems has been provided by the famous professor of business administration, ***West Churchman*** (1971). According to him, the characteristics of a system are the following:

- It is teleological (purposeful).

- Its performance can be determined.

- It has a user or users.

- It has parts (components) that in and of themselves have purpose.

- It is embedded in an environment.

- It includes a decision maker who is internal to the system and who can change the performance of the parts.

- There is a designer who is concerned with the structure of the system and whose conceptualization of the system can direct the actions of the decision maker and ultimately affect the end result of the actions of the entire system.

- The designer's purpose is to change a system so as to maximize its value to the user.

- The designer ensures that the system is stable to the extent that he or she knows its structure and function.

Churchman's concept of a designer may of course be interpreted in a religious or philosophical way (Churchman is a deeply religious scientist). A more common interpretation is, however, to see the designer as the human creator of the specific system in question (e.g. a computerized system for booking opera tickets).

Today, there is near total agreement on which properties together comprise a general systems theory. ***Ludvig von Bertalanffy*** (1955), ***Joseph Litterer*** (1969) and other distinguished persons of the systems movement have formulated the hallmarks of such a theory. The list below sums up their efforts.

- **Interrelationship and interdependence of objects and their attributes** Unrelated and independent elements can never constitute a system.

- **Holism** Holistic properties not possible to detect by analysis should be possible to define in the system.

- **Goal seeking** Systemic interaction must result in some goal or final state to be reached or some equilibrium point being approached.

- **Transformation process** All systems, if they are to attain their goal, must transform inputs into outputs. In living systems this transformation is mainly of a cyclical nature.

- **Inputs and outputs** In a closed system the inputs are determined once and for all; in an open system additional inputs are admitted from its environment.

- **Entropy** This is the amount of disorder or randomness present in any system. All non-living systems tend toward disorder; left alone they will eventually lose all motion and degenerate into an inert mass. When this permanent stage is reached and no events occur, maximum entropy is attained. A living system can, for a finite time, avert this unalterable process by importing energy from its environment. It is then said to create negentropy, something which is characteristic of all kinds of life.

- **Regulation** The interrelated objects constituting the system must be regulated in some fashion so that its goals can be realized. Regulation implies that necessary deviations will be detected and corrected. Feedback is therefore a requisite of effective control. Typical of surviving open systems is a stable state of dynamic equilibrium.

- **Hierarchy** Systems are generally complex wholes made up of smaller subsystems. This nesting of systems within other systems is what is implied by hierarchy.

- **Differentiation** In complex systems, specialized units perform specialized functions. This is a characteristic of all complex systems and may also be called specialization or division of labour.

- **Equifinality** and **multifinality** Open systems have equally valid alternative ways of attaining the same objectives (divergence) or, from a given initial state, obtain different, and mutually exclusive, objectives (convergence).

The application of these standards to the theories introduced in Chapter 3 will demonstrate that the different theories are more or less general in scope. Most of them are in fact systems theories, albeit related to a certain area of interest.

General Systems Theory is a part of the **systems paradigm** which complements the traditional scientific paradigm (see p. 11) with a kind of thinking that is better suited to the biological and behavioural realms. The objective attitude of the scientific paradigm is supplemented with *intervention*, activism and *participation* (often objectivity communicates less than subjectivity). This more comprehensive systems paradigm attempts to deal with processes such as life, death, birth, evolution, adaptation, learning, motivation and interaction (van Gigch 1992). It will also attend to explanations, values, beliefs and sentiments, that is, to consider the emotional, mental, and intuitive components of our being as realities. Consequently, the scientist becomes *involved* and is allowed to show *empathy*.

Also related to General Systems Theory is the **evolutionary paradigm** (***R. Fivaz*** 1989). Spontaneous general evolution from the uncomplicated to the complex is universal; simple systems become differentiated and integrated, both within the system and with the environment outside of the system. From elementary particles, via atoms, molecules, living cells, multicellular organisms, plants, animals, and human beings evolution reaches society and culture. Interpreted in terms of consciousness, the evolutionary paradigm implies that all matter in the universe – starting with the elementary particle – move up in levels of consciousness under the forces of evolution. The evolution *per se* thus points in a direction from the physical to the psychical. This view has many applications in sciences and makes it possible to unify knowledge from separate disciplines.

Inasmuch as scientists in the disciplines of physics, biology, psychology, sociology and philosophy all employ some mode of related thinking, a common language of concepts and terms has been established. This language embraces the common underlying principles of widely separated phenomena. Innovative and useful constructs within one area have spread to others and then merged into elements of General Systems Theory, which therefore can be defined as a *metatheory*.

On the following pages the most essential terms – those related to the general properties of systems regardless of their physical nature – are presented. These terms refer more to organization and function than to the nature of the mechanism involved. To understand them is to be familiar with the basic foundation of General Systems Theory, to possess the conceptual tools necessary to apply systems thinking to real-world systems.

Finally, the characterization of General Systems Theory made by its originator, von Bertalanffy (1967), is worth quoting:

> 'It is the beauty of systems theory that it is psycho-physically neutral, that is, its concepts and models can be applied to both material and nonmaterial phenomena.'

GST and concepts defining systems properties

First we have to define the word **system** and emphasize its subjective nature. A system is not something presented to the observer, it is something to be recognized by him. Most often the word does not refer to existing things in the real world but rather to a way of organizing our thoughts about the real world. The *constructivist* view of reality (E. von Glaserfeld 1990) states that systems do not exist in the real world independent of the human mind; only a worm's eye view defines the cell (or whatever subunit of a system) instead of a wholeness. The *fictionalist* view takes a further step and states that the systemic concept can be well suited to its purpose even if we know that it is incorrect or full of contradictions in a specific situation.

An apposite definition of the word system has been given by the well-known biologist Paul Weiss: 'A system is anything unitary enough to deserve a name.' More aphoristic is Kenneth Boulding's (1985) 'A system is anything that is not chaos', while West Churchman's view that a system is 'a structure that has organized components' seems more stringent.

An often used common-sense definition is the following: 'A system is a set of interacting units or elements that form an integrated whole intended to perform some function.' Reduced to everyday language we can express it as any structure that exhibits *order*, *pattern* and *purpose*. This in turn implies some constancy over time.

Another pragmatic definition used especially in the realm of management is that a system is the organized collection of men, machines and material required to accomplish a specific purpose and tied together by communication links.

A more scientific definition has been given by ***Russell Ackoff*** (1981), who says that a system is a set of two or more elements that satisfies the following three conditions.

- The behaviour of each element has an effect on the behaviour of the whole.
- The behaviour of the elements and their effects on the whole are interdependent.
- However subgroups of the elements are formed, all have an effect on the behaviour of the whole, but none has an independent effect on it.

An often applied mathematical definition of the word system comes from ***George Klir*** (1991). His formula is however extremely general and has therefore both weaknesses and strengths. See Figure 2:1.

In the formula, **T** stands for a set having arbitrary elements, but it may also represent a power set. **R** stands for every relationship that may be defined on the set with its special characteristics.

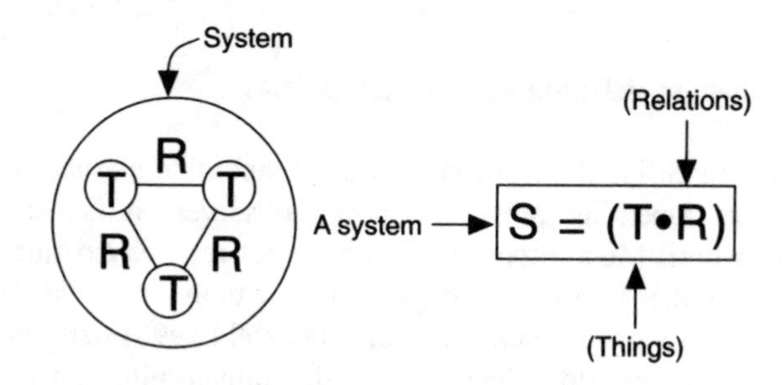

Figure 2:1 A formula defining a system.

It must however be emphasized that a set of elements, all of which do the same thing, forms an *aggregate*, not a system. To conform with the definition of a system, there has to be a functional division and coordination of labour among the parts. This implies that the components have to be assembled in a certain way in order to build a system. A system is distinguished from its parts by its *organization*. Thus, a random assembly of elements constitutes only a structure-less mass unable to accomplish anything. Nor does an orderly assembly of elements necessarily form a system. The beautiful organization of the atoms of a crystal does not qualify it to be a system; it is an end product in itself, one not performing any function.

To qualify for the name system, two conditions apart from organization have to be present: *continuity of identity* and *goal directedness*. Something that is not able to preserve its structure amid change is never recognized as a system. Goal directedness is simply the existence of a function.

Systems are usually classified as concrete, conceptual, abstract or unperceivable. The most common, the **concrete system** (sometimes called physical system), exists in the physical reality of space and time and is defined as consisting of at least two units or objects. Concrete systems can be *non-living* or *living*. Another distinction can be made, that between *natural* systems (coming into being by natural processes) and *man-made* systems.

A **living** or **organic system** is subject to the principles of natural selection and is characterized by its thermodynamic disequilibrium. As a complex, organized, and open system it is also defined by its capacity for *autopoiesis* (***H. Maturana*** and ***V. Varela*** 1974), which means 'self-renewing' and allows living systems to be autonomous. The activities of autonomous systems are mainly directed inward, with the sole aim of preserving the autonomy *per se*. Maintaining internal order or own identity under new conditions demands frequent internal reorganization.

Characteristic for autopoietic systems is *metabolism, repair, growth* and *replication*. These systems maintain their organization through a network of component-producing processes which in turn generate the same network which produced them. Advanced autopoietic systems are capable not only of organizing themselves but also of ordering their environment with growing efficiency. In contrast, an *allopoietic* system gives rise to a system which is different from itself.

The following specific qualities differentiate living systems from non-living ones.

- the presence of both *structural* and *genetic* connections in the system;

- the presence of both *coordination* and *subordination* in the system;

- the presence of a unique *control mechanism* (e.g. the central nervous system) working in a *probabilistic* manner possessing a certain number of *degrees of freedom* in the system;

- the presence of processes which *qualitatively transform* the parts together with the whole and *continuously renew* the elements.

Living systems in general are energy transducers which use information to perform more efficiently, converting one form of energy into another, and converting energy into information. Higher levels of living systems include artefacts and mentefacts. An *artefact* is an artificial object and can be everything from a bird's nest to computers and communication networks. A *mentefact* is a mental creation, exemplified here by data, information or a message. Among artefacts, a distinction has to be made between machines, tools, etc., and structures. The former have limited lives, become worn out and are replaced by better ones. The structures, however, are constructed to be permanent like the pyramids of Egypt or the Great Wall of China.

Living systems theory, formulated within the General Living Systems theory, or GLS, must be regarded as a component of the General Systems Theory. GLS, pioneered by ***James Miller*** (*Living Systems* 1976), is presented among the cornerstone theories of the next chapter.

A **conceptual system** is composed of an organization of ideas expressed in symbolic form. Its units may be words, numbers or other symbols. A conceptual system can only exist within some form of concrete system, for example a computer. An example is a computer which drafts the specifications and plans for a another physical system before it is actually created. A conceptual system (the Ten Commandments) can also regulate the operation of a physical system (the human being). In an **abstract system** all elements must be concepts. The relationships between the mental abstractions and their classes constitute the system. In psychology, for example, the structures of psychic processes are described by means of a system of abstract concepts. In an **unperceivable system** the many parts and complicated interrelationships between them hide the actual structure of the system.

All systems have a certain type of structure. Concrete systems.for example exist physically in space and time, building a specific pattern. Conceptual and abstract systems have no defined position in space nor a well-defined duration in time. However, as time goes on all systems change to a certain extent. This is called *process*; if the change is irreversible the process is called *historical*.

Here a further distinction must be made, between open and closed systems. An **open system** (all living systems) is always dependent upon an environment with which it can exchange matter, energy and information. Its main characteristic is its organization which is controlled by information and fuelled by some form of energy. The **closed system** (e.g. the biosphere) is open for input of energy only. The differences between open and closed systems are relative. An organism is a typical example of an open system but, taken together with its environment, it may be considered as a closed system.

Expressed in terms of entropy, open systems are negentropic, that is, tend toward a more elaborate structure. As open systems, organisms which are in a steady state are capable of working for a long time by use of the constant input of matter and energy. Closed systems, however, increase their entropy, tend to run down and can therefore be called 'dying systems'. When reaching a state of equilibrium the closed system is not capable of performing any work.

An **isolated system** then is one with a completely locked boundary closed to all kinds of input. Independent of its structure or kind, it is constantly increasing its entropy into a final state of genuine equilibrium. While this concept is very seldom applicable in the real world, the cosmos is the environmentless, isolated system context for all the other systems which may arise within it.

The systems that we are interested in exist within an *environment*. The immediate environment is the next higher system minus the system itself. The entire environment includes this plus all systems at higher levels which contain it. Environment may also be defined as both that which is outside of the direct control of the system and any phenomenon influencing the processes and behaviour of the system. For living systems, however, environment must be seen as a part of the organism itself. The internal structure of living organisms contains elements which earlier in the evolution were part of its external environment. This is confirmed, among other similarities, by the chemical composition of blood and sea water.

Environment is something which exists in **a space**, a concept which is defined with respect to the kind of system in focus. *Pragmatic space* is that of physical action which integrates a living system with its natural, organic environment. *Perceptual space* is that of immediate orientation, essential for the identity of a conscious being. *Existential space* forms a stable image of an individual environment and connects it to a social and cultural identity. *Cognitive space* is the conscious experience of the physical world, while logical or *abstract space* belongs to the world of abstract or conceptual systems, thus providing us with a tool to describe the others.

Through the constant interaction between system and environment, environment affects systems and systems in turn affect the environment. When it comes to social systems this interaction is especially pronounced. Its scope is suggested in the following pairs.

Living system	Environment
– Society	– Nature
– We	– Them
– Self	– The other
– Ego	– Id
– Mind	– Body
– Consciousness	– Subconsciousness

In order to define a system's environment its *boundary* must be defined. The boundary surrounds the system in such a way that the intensity of interactions across this line is less than that occurring within the system. Often a non-spatial marker, it denotes what does or does not belong to the system. To cross a boundary normally requires modification or transformation in some way. In the case of information, boundaries possess a coding and decoding property. In other words that which comes out is very seldom identical with that which goes into a system. As systems do not always exist boundary to boundary, the concept of *interface* is necessary to denote the area between the boundaries of systems.

As rare as the concept of a closed system is that of a solitarily existing open system. Generally, systems are part of other systems and are included in a **hierarchy of systems**. Systems theory regards the concept of hierarchy as a universal principle existing in inorganic nature, in organic and social life and in the cosmos. In a hierarchic structure, subsets of a whole are ranked regressively as smaller or less complex units until the lowest level is reached. The lowest level *elements* build *subsystems* that in turn structure the system, which itself is a part of a superior *suprasystem*. The ranking of these is relative rather than absolute. That is, the same object may be regarded as an element, a system or a component of the environment, depending on the chosen frame of reference. See Figure 2:2.

Hierarchical thinking creates what has been called the paradox of hierarchy. It implies that a system can be described if regarded as an element of a larger system. Presenting a given system as an element of a larger system can only be done if this system is described as a system.

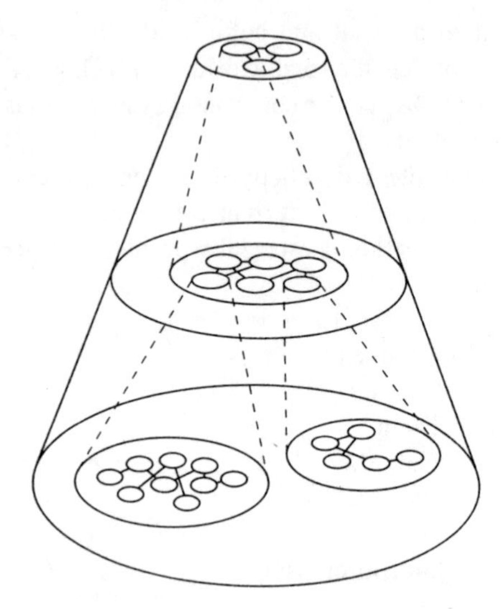

Figure 2:2 A multilevel systems hierarchy.

A more elaborate hierarchical terminology used in this context is:

- macrosystem
- system
- subsystem
- module
- component
- unit
- part

At a given level of the hierarchy, a given system may be seen as being on the outside of systems below it, and as being on the inside of systems above it. A system thus has both *endogenous* and *exogenous* properties, existing within the system and determined outside of the system respectively. Again, as above, the status of a component in a system is not absolute: it may be regarded as a subsystem, a system or an element of the environment. In order to carry out their functions in a suprasystem, subsystems must retain their identities and maintain a certain degree of autonomy. A process whereby the interaction increases in a certain part of the system often ends up in a new local structure. This is called *centralization* and small variations within this part can produce essential changes of the whole system.

Another kind of hierarchic view is expressed in the **holon** (from wholeness) concept, coined by the Hungarian-born author, ***Arthur Koestler***, in 1967. Wholes and parts do not have separate existences in living organisms or social organizations. These systems show both cohesion and differentiation. Their integrative and self-assertive tendencies exist side by side and are reflected in their cooperative behaviour. This 'Janus' effect (from the Roman two-faced god Janus) is a fundamental characteristic of subwholes in all kinds of hierarchies. The global structure of the holon hierarchy is nested. At least five levels are discernible in Figure 2:3.

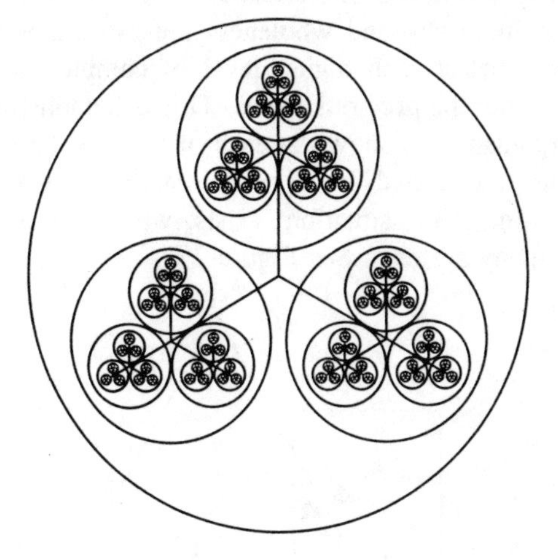

Figure 2:3 Integrative and assertive relationships of a holon represented by circles.

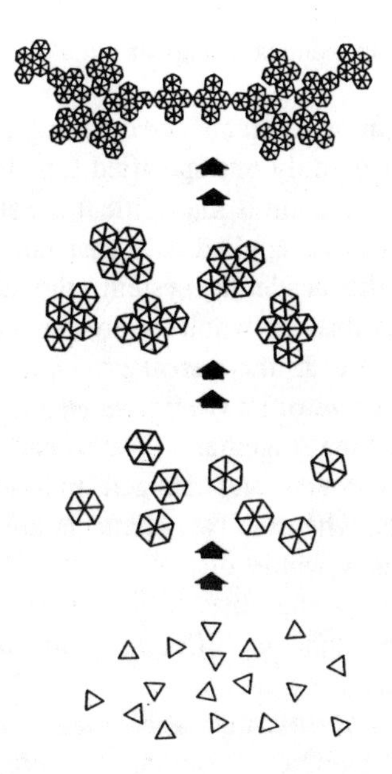

Figure 2:4 An organism regarded as the wholeness of organismic symbiosis.

Normally, the term *wholeness* applied to a system indicates the following: variation in any element affects all the others bringing about variation in the whole system. Likewise, variations of any element depend upon all other elements of the system. In a sense, there is a paradox of wholeness telling us that it is impossible to become conscious of a system as a wholeness without analyzing its parts (thereby losing the wholeness).

The concepts of hierarchy and wholeness are especially relevant in living things where organisms at each higher level of complexity originate as as a symbiosis of those from the previous levels. This is demonstrated in Figure 2:4 where different organisms are shown at each of the four levels.

Systems can be interrelated in a non-hierarchical way when part of a *multilateral* structure. This situation exists when certain elements occur simultaneously in many systems. See Figure 2:5.

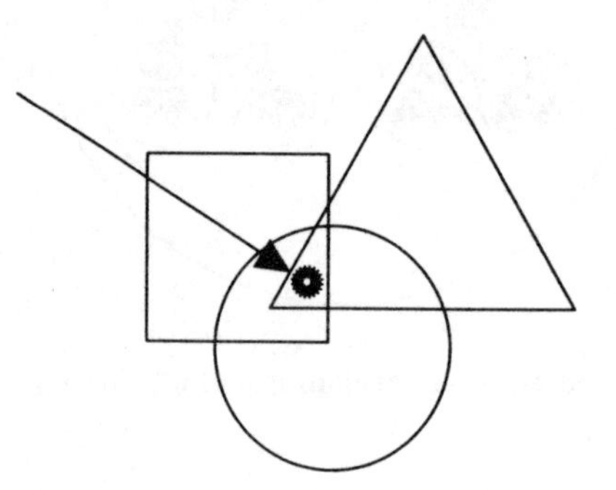

Figure 2:5 System element as part of a multilateral structure.

In a system, elements can be interconnected during a certain period of time. If the connection exists during only one specified time the multilateral structure is called *temporal*. If the connection is intermittent the structure is called *cyclic*.

The concept of system can be applied to a vast number of different phenomena: the solar system, the academic system, the nervous system, etc. A characteristic of them all is that the whole is greater than the sum of its parts, a phenomenon often referred to as the *system principle*. This principle includes the system's *emergent properties* or its *synergetic effects*. Synergetic comes from the Greek word for 'working together'. Water can illustrate an emergent phenomenon: although hydrogen and oxygen individually have no water-qualities, water will emerge when the two elements are brought together.

A suprasystem taken as a whole displays greater behavioural variety and options than its component systems, that is, it is more *synergetic*. Each system has a special organization that is different from the organization of its components taken separately.

Normally systems show *stability*, that is, constancy of structure and function under fluctuation, which maintains the same internal state under various pressures. Systems which can restore their stability by changing operation rules when important variables exceed their limits are said to be *ultra-stable*. Stability then does not exclude *adaptability*; only systems which change with time and adjust to environmental pressures can survive.

In open systems, for example, biological and social systems, final states or objectives may be reached in different ways and from disparate starting points. This property of finding equally valid ways is called *equifinality*. The reverse condition, achievement of different ends through use of the same means, is called *multifinality*.

A basic concept in GST is that of **entropy**. Originally imported from the area of thermodynamics, it is defined as the energy not available for work after its transformation from one form to another (see also p. 12). Applied to systems it is defined as a measure of the relative degree of disorder that exists within a closed system at a defined moment of time. The natural tendency of physical objects to disintegrate and fall into random distribution can be studied in a sand castle built on the beach on a hot day. How biological organisms deteriorate as they age can be seen everywhere in our environment.

Both examples relate to the effects of entropy, of the transformation of energy from high quality to low. Living systems can however, as open systems, counteract this tendency through purpose and organization, by importing more energy from their environment than they expend to it. Storing the surplus energy in order to survive is to reverse the entropic process or to create *negentropy*. A living being can only resist the degradation of its own structure. The entropic process influencing the structure and environment of the whole system is beyond individual control.

Systems may be classified according to type of complexity, as has been done by ***Warren Weaver*** (1968).

In the **organized-complexity system,** the typical form for living systems, only a finite but large number of components will define the system. Systems within this category can also be classified as middle-number systems. When a limit is reached the living system decomposes into irreducible particles. As stated earlier, the total system always represents more than the sum of its parts. This type of complexity cannot be treated with the statistical techniques that so effectively describe average behaviour within unorganized complexity. Successful investigations of organized complexity became feasible first with the emergence of computer technology.

The **unorganized-complexity system** can only refer to non-living systems where the number of variables is very large and in which each variable has a totally unpredictable or unknown behaviour. The system has nevertheless orderly average properties and may be defined in terms of a probability distribution according to an infinite number of events. Its behaviour can be explained by the laws of statistical mechanics and its components may form aggregates. The frequency and type of telephone calls in a large telephone exchange offer a good example.

The **organized-simplicity system** is characterized by simple systems such as machines and other human artefacts having only a small number of components. This kind of system may be treated analytically.

A similar classification of systems has been made by ***Herbert Simon*** (1968). He distinguishes decomposable, nearly decomposable and non-decomposable systems. In a **decomposable system** the subsystems can be regarded as independent of one another. A given example is helium, an inert gas: the intermolecular forces will be negligible when compared to the intramolecular forces. In **near decomposable systems** the interaction between the subsystems is weak but not negligible. The intercomponent interactions are usually weaker than the intracomponent interaction. Organizations may be considered to be near decomposable. **Non-decomposable systems** are directly dependent on other systems or explicitly affect them. A heart/lung machine is such a system.

Another classification of systems is made on the basis of their behaviour or function. A classification of this kind has been made by the doyen of management research, ***Russell Ackoff*** (Ackoff 1971). According to this, **goal-maintaining systems** attempt to fulfil a predetermined goal. If something deviates, there is only one response (conditional) to correct it. Here the thermostat and other simple regulatory mechanisms can serve as examples.

In **goal-seeking systems** choices concerning how to deal with variable behaviour in the system are possible. Previous behaviour stored in a simple memory permits changes based on learning. The autopilot meets the requirements: it maintains a preset course, altitude and speed.

Multigoal-seeking systems are capable of choosing from an internal repertoire of actions in response to changed external conditions. Such automatic goal changing demands distinct alternatives; generally the system decides which means of achievement are best. A prerequisite is an extended memory with the ability to store and retrieve information. The automatic telephone exchange is a good example.

Reflective, goal-changing systems reflect upon decisions made. Information collected and stored in the memory is examined for the creation of new alternatives for action. Will, purpose, autonomy, feedforward (see p. 48), learning and consciousness define this process, existing only within living systems.

Another often used system dichotomy is that of static and dynamic systems. A **static system** is a structure which is not in itself performing any kind of activity. A **dynamic system** has both structural components and activity. Examples of such systems are respectively a radio tower and a military squad with its men, equipment and orders.

Some other special categories of systems which need to be mentioned are the irrational and null. Both violate the principle of causality (see p. 9) and cannot be handled by way of rational judgement. In the **irrational system** there is no correspondence between input and the presumed system response. In the **null system**, all input produces no output or an output is produced without significant input. While both systems are also unmeasurable systems, we must first be aware of the difficulties often involved in identifying complex system flows. 'Occult behaviour' sometimes has a very natural cause.

Sometimes it is necessary to apply some basic mathematical criteria to the concept of systems. For a **continuous system** the input can be continuously and arbitrarily changed; the result will then be a continuous variable output. In a **discrete system** the input is changed in discrete steps, giving corresponding discrete changes in the output.

It may also be necessary to distinguish between deterministic and stochastic systems. According to the principle of nature's predictability (see p. 10) the **deterministic system** has inputs and outputs capable of being interpreted and measured on a single-event basis. The output will be the same for each identical input; repeated trials will always give the same results. The **stochastic system** works instead with identical inputs and its elements cannot be returned to their original state. The factors influencing the system obey uncertainty and statistical variation. Nonetheless, if appropriate stochastic analysis is employed, systemic behaviour may be possible to predict.

Finally, a distinction has to be made between simulative and non-simulative systems. Extremely small changes in the input into systems which are large-scale, complex and nonlinear are often amplified through positive feedback. Such changes can thus initiate exponential transformations of the whole system. An example of a **non-simulative system** is global weather, characterized by *deterministic chaos*. The system sensitivity for initial data eludes prediction. Furthermore, *any* physical system that behaves in a non-periodic way is unpredictable.

The 'butterfly effect' where the flaps of the wings of a butterfly start a movement in the air which ends up as a hurricane has fascinated many and captures the unpredictability of non-linear systems. No computer program exists which can model this system. Such a program would be just as complex as the weather system itself. Therefore, some meteorologists say that the only computer capable of simulating the global weather is the Earth itself, the ultimate analogue biocomputer.

In a **simulative system** the complexity of the computer program always falls far below the complexity of the system simulated.

Cybernetics and concepts defining systems processes

In order to predict the behaviour of a rational system before a certain response from it occurs, it is essential to have some knowledge of general control mechanisms. Although automatic control systems have been documented in the Western field of engineering for some 2000 years, the working theory of these has been limited and seldom used outside of engineering. (The Greek Ktesibios invented in 300 B.C. an automatic control device: the water flow that controlled his water clock, or *clepsydra*.)

In his book dating from 1948, *Cybernetics or Control and Communication in the Animal and the Machine*, **Norbert Wiener**, an American researcher at MIT, gave control theory new life. As a mathematician and universal thinker his fascination for logic and electricity intertwined with his insight in automation led to the ideas of cybernetics. The term **cybernetics** is derived from the Greek noun, *kubernetes,* which associates to pilot or rudder.

One of the oldest automatic control systems is in fact related to the turning of a heavy ship's rudders: the steam-powered steering engine. This supportive mechanism was called a *servo* (here from the Latin *servitudo* from which English has its servitude and slave). In his book Wiener intended cybernetics to embrace universal principles relevant for both engineering and living systems. (One of these principles is that all the processes in the universe seem to be cyclic.) He did succeed in showing that these principles could fruitfully be applied at the theoretical level in all systems. Shortly after cybernetics emerged as an independent area of its own, it became part of GST. For practical purposes the two areas were integrated within the wide domain of different problems that became the concern of systems science.

In cybernetics, the concepts of control and communication are closely interrelated. Information concerning function and control is communicated among the parts of a system but also between it and its environment. This has the aim of achieving a condition of **equilibrium**. In living systems, this holding of physiological variables within certain limits is called **homeostasis**. Cybernetics, then, concerns the restoring of stability within all kinds of systems.

The fact that cybernetic control systems operate with a low, often insignificant, expenditure of energy means that their degree of efficiency is high. This is possible inasmuch as their basic function is to process information, not to transform energy. Cybernetic regulation must not be confused with any amplification of the affected flow which may happen when amplification exists as well.

As a starting point for the comprehension of the basic terms of cybernetics a system may be represented by three boxes: the black, the grey and the white. The purposeful action performed by the box is its *function*. Inside each box there are *structural components*, the static parts, *operating components* which perform the processing, and *flow components,* the matter/energy or information being processed.

Relationships between the mutually dependent components are said to be of *first order*. Here the main example is symbiosis, the vitally important cooperation between two organisms. Relationship of *second order* is that which adds to system performance in a synergistic manner. Relationship of *third order* applies when seemingly redundant duplicate components exist in order to secure a continued system function.

Each box contains processes of **input, transformation** and **output**. (Note that output can be of two kinds: products useful for the suprasystem and/or waste.

Also, note that the input to one system may be the output of its subsystem.) Taken together these processes are called **throughput**, to avoid focus on individual parts of internal processes.

The box colours denote different degrees of user interest in the understanding or knowledge of the internal working process of a system. A **black box** is a primitive something that behaves in a certain way without giving any clue to the observer how exactly the result is obtained. As Kenneth Boulding wrote:

> A system is a big black box
> Of which we can't unlock the locks
> And all we can find out about
> Is what goes in and what comes out.

A black-box approach can therefore be the effective use of a machine by adjusting its input for maximum output (cold shower to bring down fever). A **grey box** offers partial knowledge of selected internal processes (visit nurse for palliative treatment). The **white box** represents a wholly transparent view, giving full information about internal processes (hospitalize for intensive treatment). This command of total information is seldom possible or even desirable.

Below a certain level questions cannot be answered, or posed; complete information about the state of the system can therefore not be acquired. See Figure 2:6.

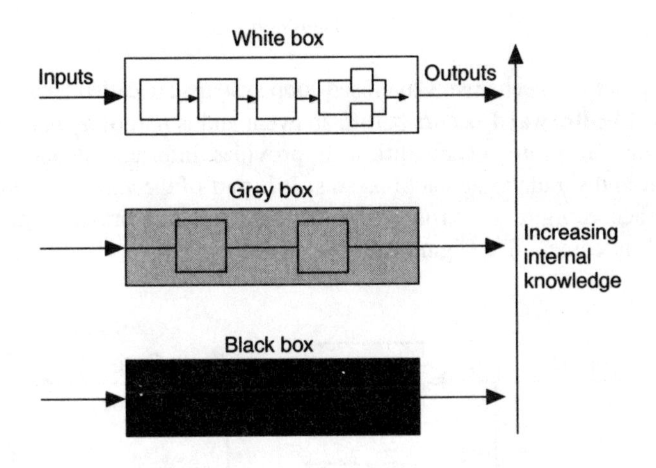

Figure 2:6 Degrees of internal understanding.

However, when good understanding of the whole transformation process is necessary, the following five elements have to be calculated.

- **The set of inputs** These are the variable parameters observed to affect the system behaviour.

- **The set of outputs** These are the observed parameters affecting the relationship between the system and its environment.

- **The set of states** These are internal parameters of the system which determine the relationship between input and output.

- **The state-transition function** This will decide how the state changes when various inputs are fed into the system.

- **The output function** This will decide the resulting system output with a given input in a given state.

System processes may or may not be self-regulated. A self-regulated system is called a **closed-loop system** and has its output coupled to its input. In the **open-loop system**, the output is not connected to its input for measurement. An example is an automatic sprinkler system, depicted as an open-loop system in Figure 2:7.

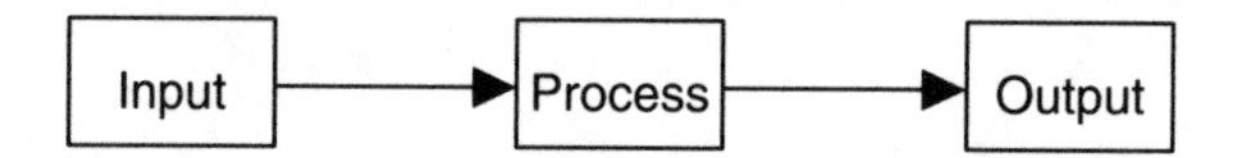

Figure 2:7 Open-loop system.

The regulatory mechanisms of closed-loop systems are called feedforward and feedback. **Feedforward** occurs before an event and is part of a *planning loop* in preparation for future eventualities. It provides information about expected behaviour and simulates actual processes. It is part of the anticipatory control of a system according to an existing internal model relating present inputs to their predicted outcomes. See Figure 2:8.

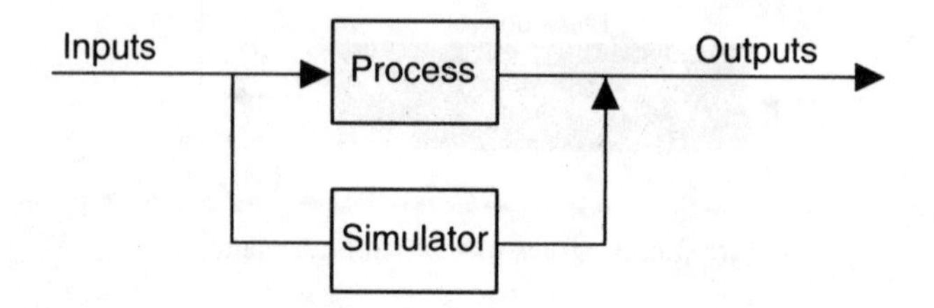

Figure 2:8 Feedforward loop.

Feedback is a basic strategy which allows a system to compensate for unexpected disturbances and is often defined as the 'transmission of a signal from a later to an earlier stage'. Information concerning the result of own actions is thus delivered as a part of information for continuous action. As a control mechanism it acts on the basis of its actual rather than its expected performance. Feedback is a key concept in cybernetics. A metaphysical limerick has been dedicated to it by an anonymous poet.

Said a fisherman at Nice,
'The way we began was like this
A long way indeed back
In chaos rode Feedback
And Adam and Eve had a piece.'

A generalized theory has been developed to describe the behaviour of closed-loop systems and of systems containing a number of interacting elements using feedback.

System conduct may however become very complex if several feedback elements are interconnected; the resulting dynamics will often be difficult to calculate. The main concepts of this generalized theory are presented in the sections below.

Negative feedback is a fraction of the output delivered back to the input, regulating the new output to a multiplier smaller than one. This kind of feedback tends to oppose what the system is already doing, and is thus negative. An increase on the feedback level generates a decrease in the output, providing self-correction and stabilization of the system. Systems with feedback automatically compensate for disturbing forces *not necessarily* known beforehand. The principle of the negative feedback loop is seen in Figure 2:9.

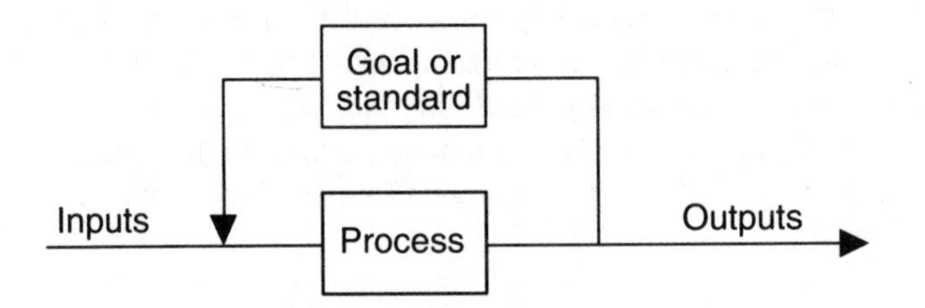

Figure 2:9 Feedback loop.

A device which acts *continuously* on the basis of information in order to attain a specified goal during changes, called a **servomechanism**, is an example of applied negative feedback. Its minimal internal structure consists of a *sensor*, an *effector* and a connecting link. A simple servomechanism is ***James Watt***'s centrifugal regulator from the 18th century. See Figure 2:10.

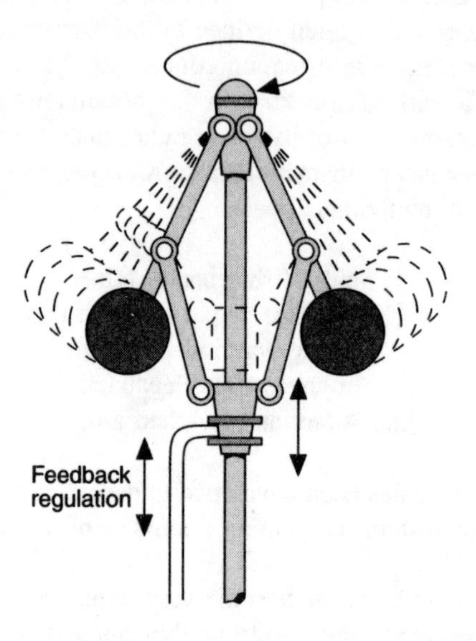

Figure 2:10 James Watt's speed controlling centrifugal regulator. Engine speed change generates counteracting forces from the regulator. The steam is choked or released, thereby returning the engine to normal operating speed.

The perfect servomechanism corrects errors before they occur. Its smooth and coordinating activity is dependent upon the amount of compensatory feedback. Both under- and over-compensation generate oscillations that are more or less harmful to the regulated system. Another example is the simple but reliable pneumatic autopilot in the DC-3 aircraft. Corrections within predefined settings (altitude, course) are handled by the system while changes of the system itself (new course, etc.) are determined by its suprasystem, here the pilot. Directions (route, schedule) are given by the supra-suprasystem, the flight traffic control.

A control mechanism can also be *discontinuous*. An example is the simple thermostat which can only perform two actions: turn the heat on or turn it off. Discrete control of this kind is common in all kinds of modern electronic equipment.

If the multiplier is greater than one, a state of **positive feedback** exists. Here each new output is larger than the previous, with an exponential growth and a deviation-amplifying effect. A positive feedback mechanism is always a 'run-away' and temporary phenomenon. Its self-accelerating loop is normally brought to a halt before the process 'explodes' and destroys the system. A negative feedback inside or outside of the system will sooner or later restore more normal behaviour. See diagrams in Figure 2:11.

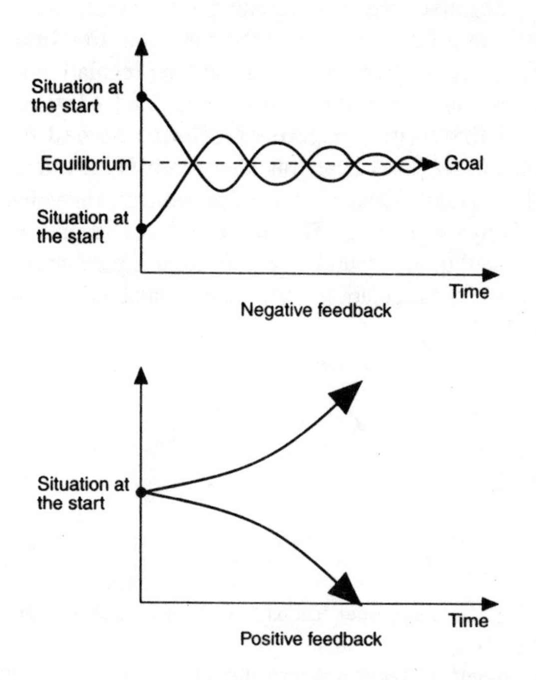

Figure 2:11 The nature of negative and positive feedback.

The combined effects of an emerging positive feedback being inhibited by a growing amount of negative feedback most often follow the nonlinear logistic equation which exhibits sigmoid growth. The effect of a shift in loop dominance in a population-growth diagram is shown in Figure 2:12. The loops change when the population reaches half of its maximum.

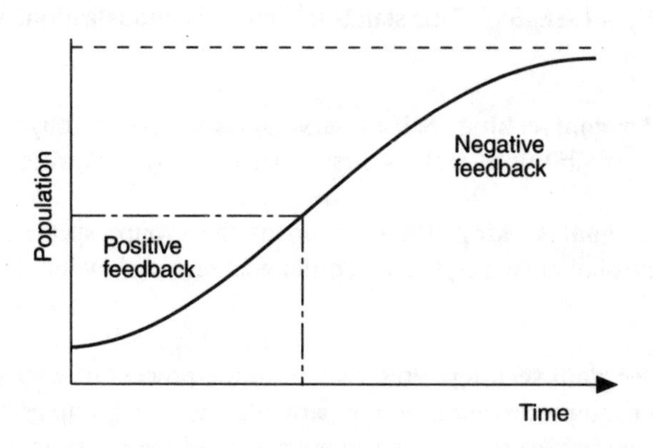

Figure 2:12 Shifting of loop dominance in population-growth diagram.

The elementary negative feedback presented here operates according to a preset goal. The only possibility is to correct the deviation. Conditional response is impossible inasmuch as no alternative exists and the regulation normally works exponentially toward the equilibrium state. This kind of direct deterministic regulation is called **first order, negative feedback. Second order, negative feedback** is defined as feedback based on other feedback. It is thus more indirect than that of the first order, which comes either from an immediately preceding step or directly from monitoring. This more indirect second-order regulation causes sinusoidal oscillations around an equilibrium if undamped. If damped by a first order feedback, the regulation will follow a damped sinusoidal curve. See curves in Figure 2:13.

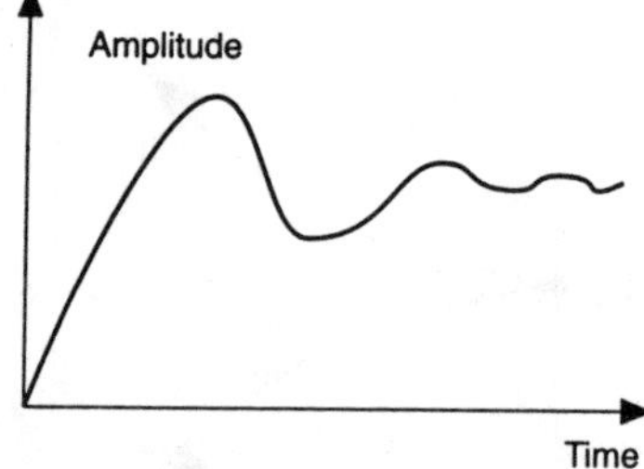

Figure 2:13 Second-order feedback with sinusoidal oscillations.

Higher-order, negative feedback regulation also operates with oscillations around an equilibrium. Over-reacting feedback chains can bring about a growing reaction amplitude, thus rendering the system unstable. To be stable, the regulatory mechanisms have to be adequately damped. The system's own friction is often enough to have this function.

Different levels of goal seeking as a cybernetic feedback process have been proposed by ***Karl Deutsch*** (1963). His goal-seeking hierarchy with four levels may be compared with Ackoff's behavioural classification of systems (see p. 44).

- **First-order goal seeking** This stands for immediate satisfaction, adjustment, reward.

- **Second-order goal seeking** Self-preservation is achieved through the preservation of the possibility of seeking first-order goals by controlling the same.

- **Third-order goal seeking** Preservation of the group, species, or system requires control over first and second order goal seeking beyond the individual life-span.

- **Fourth-order goal seeking** Preservation of the process of goal seeking has priority over the preservation of any particular goal or group as above. This is in effect the preservation of the relationships of the ecosystem.

Sometimes it is necessary to distinguish between extrinsic and intrinsic feedback. **Extrinsic feedback** exists when the output crosses the boundary and becomes modified by the environment before re-entering the system. **Intrinsic feedback** prevails when the same output is modified internally within the system boundary. While the concept of feedback is generally defined as being intrinsic, from the system's point of view, both types are equal. Normally the system is unaware of the actual feedback type. See Figure 2:14.

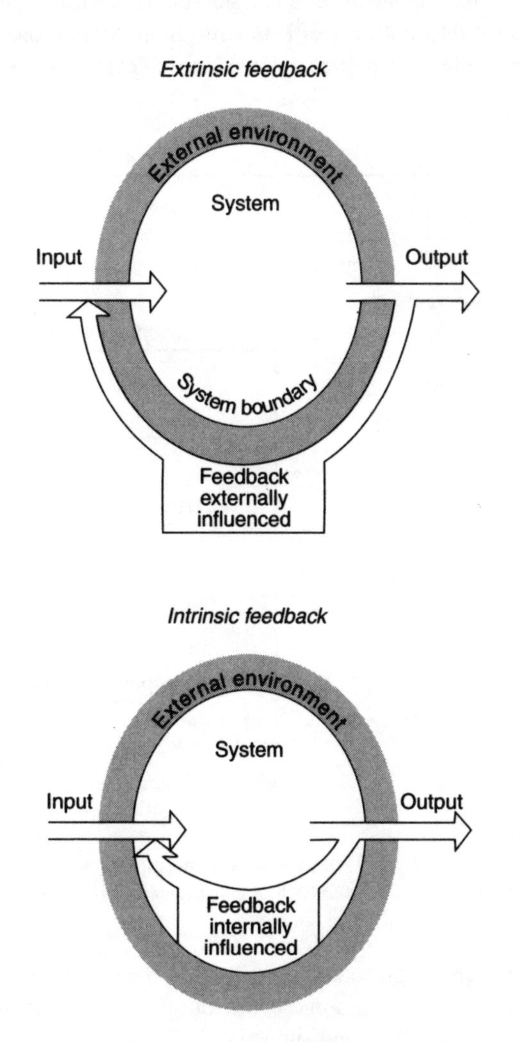

Figure 2:14 Extrinsic and intrinsic feedback.

In cybernetic control cycles, time plays an important role. Variations in speed of circulation and friction between different elements of the system can occur.

Such delays and lags are important regulatory parameters which counteract inherent oscillatory tendencies of a feedback control process. They are often employed to set physical limitations on the system, slowing down the action, but dynamically. Important variables (especially the output) are prevented by this limitation from jumping abruptly from one value to another.

A **delay** can completely inhibit a regulatory action for a certain amount of time, after which action starts with full impact. A **lag** is a gradual regulatory force, reaching its full impact after a certain amount of time. Feedback systems with lags may have destabilizing effects with the pertinent loss of control. The effects of delays and lags combined are even more devastating (see Figure 2:15).

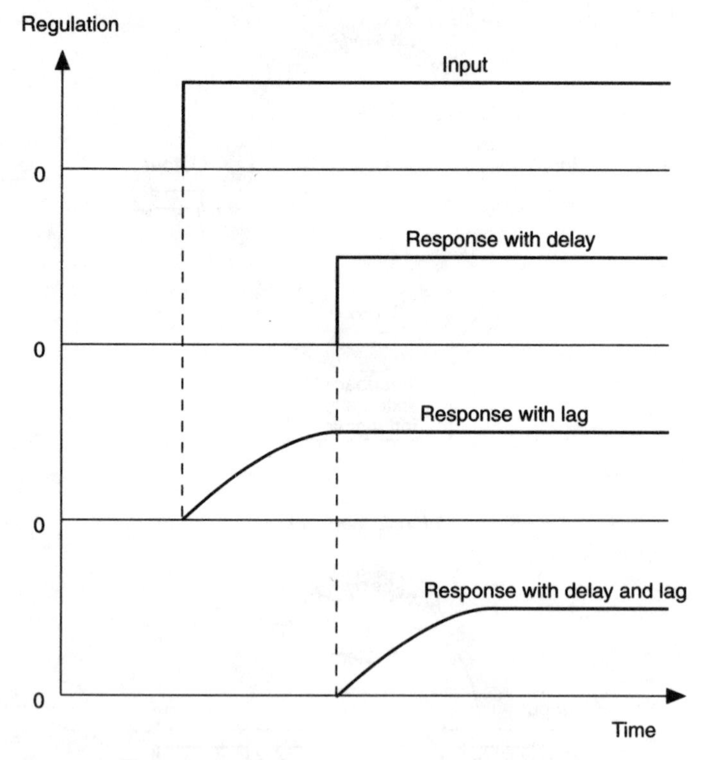

Figure 2:15 Delays and lags in control cycles.
(Reprinted with permission from J.P. van Gigch, *Applied General Systems Theory*, Harper & Row, NY, 2nd Ed., 1978.)

The feedback processes presented here operate in a variety of control systems. Their main function is to keep some behavioural variables of the main system within predefined limits. The end objective is to maintain an output that will satisfy the system requirements. The ideal control system produces a regulation which cancels out the effect of possible disturbances completely. A general control system using a basic control cycle is presented in Figure 2:16.

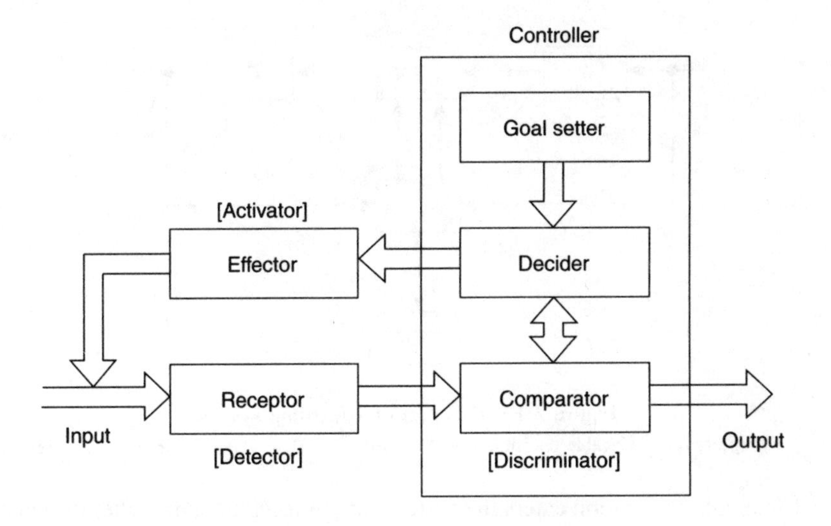

Figure 2:16 A general control system.

The first fundamental component of the regulatory mechanism in the **basic control cycle** is the *receptor*, a sensor registering the various stimuli which, after conversion into information, reach the *controller*. A comparison is made between the sensor value and a desired standard stored in the *comparator*. The difference provides a corrective message which is implemented by the *effector*. Through monitoring and response feedback to the receptor, self-regulation is achieved. Figure 2:16 shows that the regulation takes place on the input side and the sensing mechanism is situated on the output side. In more sophisticated systems with third-order feedback, the controller also includes a *goal-setter*, a *decision-maker* and possibly a *designer* which formulates both the goals and the decision rules of the system.

We have seen earlier that one of the most significant advantages of living systems was adaptation achieved by learning. This advantage is however not restricted to living systems only; machines working according to cybernetic principles may also be able to learn. If information moving backward from a system's performance is able to change the general method and pattern of performance, it is justifiable to speak of learning.

A general cybernetic pattern for a system capable of learning is shown in Figure 2:17.

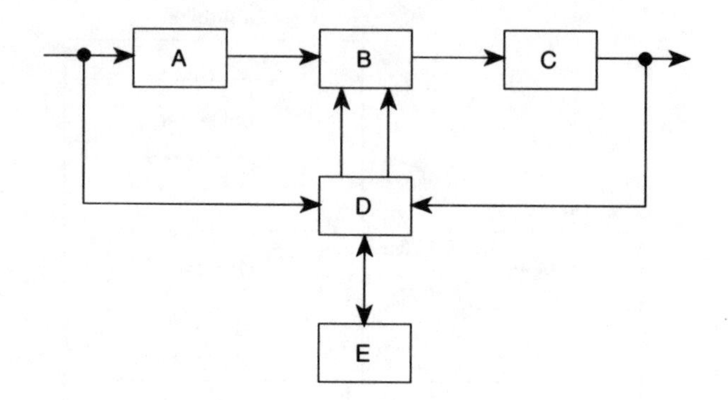

Figure 2:17 Diagram of a learning system.
A = Receptor; B = Educable decision unit; C = Effector; D = Comparator; E = Goal setter.

The input information enters the system *via* the receptor and reaches the educable internal decision unit. After processing, the information will reach the effector and there become an output. The behaviour of the decision unit is however not predetermined. (Through a double path the same input, and the output decision as well, are simultaneously led to an evaluating mechanism.)

From the evaluating mechanism, the comparator, a parallel path leads to the decider. This receives simultaneously the same input as is given to the receptor, also the same output as delivered from the effector. The decision unit compares the cause in the input with the effect of the output on the basis of the evaluation criteria stored in the comparator. If the decision is correct or 'good', the decider is 'rewarded', and if incorrect or 'bad', the decider is 'punished'. In reality this results in a modification of its internal parameters which is a kind of self-organization and learning.

There is of course a risk of confusing self-organization with learning. All systems able to learn must necessarily organize themselves, but systems can organize themselves without learning. The faculty for modifying its behaviour and adapting is not in itself sufficient for it to be regarded as a learning system. The point is that the rules must be adjusted in such a way that a successful behaviour is reinforced, whereas an unsuccessful behaviour results in modification. Thus, the important thing is the internal modification of the information transfer.

When defining living systems, the term **homeostasis** stands for the sum of all control functions creating the state of dynamic equilibrium in a healthy organism. It is the ability of the body to maintain a narrow range of internal conditions in spite of environmental changes. All systems do however age and, from a certain point of maturation, slowly deteriorate toward death. This phenomenon is called *homeokinesis* and gave rise to the concept of the **homeokinetic plateau**, depicted in the diagram of Figure 2:18.

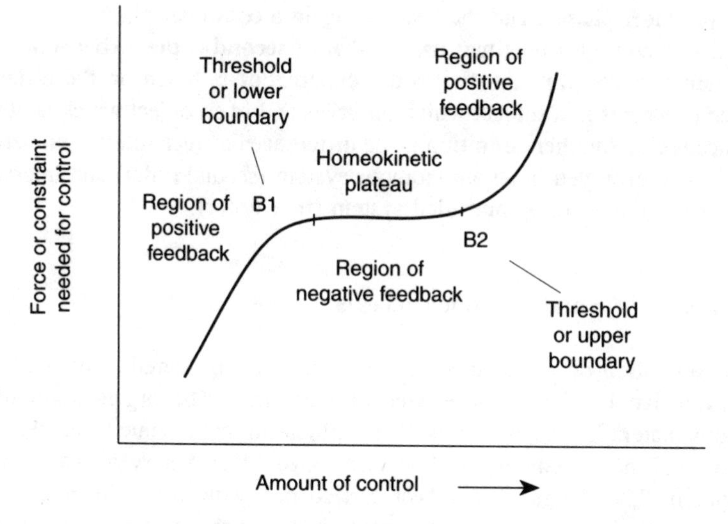

Figure 2:18 The homeokinetic plateau.

This constant deterioration can be compensated for by extended control and mobilization of resources within the limits of the homeokinetic plateau. Here the negative feedback is stronger than the positive and a temporary homeostasis can be maintained within the thresholds of the plateau. Below and above the thresholds, the net feedback is positive, leading to increased oscillations and finally to the collapse of the system. The only alternative to break-down for a system going outside the homeostatic plateau is an adaptation through change of structure. This adaptation is however beyond the capabilities of an individual organism.

The homeokinetic plateau is a quite natural part of what can be called a **system life cycle**. In living systems this consists of birth, evolution, deterioration, and death. In non-living systems, such as more advanced artefacts, the system life cycle can be divided into following phases:

- identification of needs
- system planning
- system research
- system design
- system construction
- system evaluation
- system use
- system phase-out

Note that the first phase can be considered a consumer phase, the intermediate phases producer phases, and the last two again a consumer phase.

Finally, a concept sometimes used is that of **second-order cybernetics**. The distinction between this and first-order cybernetics is based on the difference between processes in a subject which observes and in an object which is observed respectively. Another definition is the difference between interaction between observer and observed in an autonomous system (second order) and interaction among the variables of a controlled system (first order).

General scientific and systemic concepts

The accumulation of scientific knowledge may be considered to be one of the most extensive intellectual processes of humanity. The organization of the enormous material, a science in itself, is influenced by systemic principles. (See Namilov and the systems view of science, page 129.) A survey of the content in a specific knowledge area is best carried out using a top-down approach, beginning with the area's prevalent world view. For readers unfamiliar with the scientific vocabulary related to the hierarchic organization of scientific knowledge, the main concepts are presented below. According to the scientific tradition, theories should be *explicit* (not based on interpretation or intuition), *abstract* (not referring to concrete examples), and *universal* (valid in every place and at anytime). This implies that a theory regarding the behaviour of certain physical particles therefore relates to every individual particle in the universe, without exception.

A **world view** is
a grand paradigm including the beliefs and philosophical preferences of the general scientific community.

A **paradigm** is
a common way of thinking, held by the majority of members of a specific scientific community.

A **theory** is
a broad coherent assembly of systematic explanatory schemes, consisting of laws, principles, theorems and hypotheses.

A **law** is
a generalization founded on empirical evidence, well established and widely accepted over a long period of time.

A **principle** is
a generalization founded on empirical evidence but not yet qualifying for the status of a law.

A **theorem** is
a generalization proven in a formal mathematical, logical way.

A **hypothesis** is
a proposition which is intuitively and empirically considered to be true.

An **axiom** is
impossible to prove or deduce from something else, but is a starting point for the hierarchy of scientific abstractions presented here.

It is important to understand that present scientific 'truth' descends from observation and experiments. These are also the starting point for the construction of a theory which hopefully corresponds to the observations. The theory itself must be considered as an instrument to handle a formal symbolic system in order to exceed the limitations of thought. If so, this does not, however, prove its truth; it is 'only' the best we have for the moment. The truth of science is always provisional, and accordingly, the theory must be subject to change as new information appears on the horizon. The search for a better theory is a perpetual challenge for new generations of scientists.

A concept closely related to the theory is the **model**, which can be considered a link between theory and reality. To use a model is to visualize a theory or a part of it. A closer look at the model tells us that it is a phenomenon which somehow mimics or *represents* another primary entity. It may also be expressed as 'one thing we think we hope to understand in terms of another that we think we do understand' (Weinberg 1975). As a theoretical construct it fits the known, available facts into a neat and elegant package. It is an imitation or projection of the real world, based on the constructor's problem area of interest. In this simplified version of reality certain features are stereotypical. The model brings out certain characteristic features in the object of study, simultaneousely excluding others. The quality of a model can only be judged against the background of the purpose of its origin.

Models are employed to develop new knowledge, to modify existing knowledge or to give knowledge new applications. From a pedagogical point of view, models are used to render theories more intelligible. Models can also be used to interpret a natural phenomenon or to predict the outcome of actions. Through the use of models it becomes possible to know something about a process before it exists. The model can be subjected to manipulations that are too complex or dangerous to perform in full scale. Also, to use a model is less costly than direct manipulation of the system itself would be.

When a model does not work in reality this can sometimes be ascribed to the fact that the model has been confused with reality. The tool must be separate from the solution and the method from the result. Models are nevertheless in a sense indispensable as most often reality is far too complex to be understood without their help.

Models are commonly classified as iconic, analogue, symbolic, verbal and conceptual. **Iconic,** or physical, models look like the reality they are intended to represent. One example is a scale model of a ship's hull, used to collect information concerning a proposed design. Full-scale models are always iconic; they are used for the same purpose although their dimensions coincide with those of the real object. Even a living mannequin is a full-scale iconic model.

Analogue models represent important qualities of reality through similarity in relations between entities expressed in entirely different forms, that are easier to handle. Such models behave like the reality they represent without looking like it. An example is a mathematical graph or a terrain map.

Symbolic models use symbols to denote the reality of interest. Normally general and abstract, they are often more difficult to construct but easier to use than other models. Examples are mathematical, linguistic or decision-making models. A **schematic** model reduces a state or event to a diagram or chart. A circuit diagram of an electronic amplifier exemplifies a schematic model of the actual hardware. Another kind is a flow chart describing the order of events in different processes.

A **mathematical** model uses mathematical symbols to describe and explain the represented system. Normally used to predict and control, these models provide a high degree of abstraction but also of precision in their application. A warning regarding the inevitable dilemma associated with mathematical models has, however, been given by Einstein (1921) when he says: 'When mathematical propositions refer to reality they are not certain; when they are certain, they do not refer to reality.'

A **verbal** model depicts reality through the use of verbal statements that set forth the relationships between the concepts. **Conceptual** models are theoretical explanations; in accordance with their final purpose these models are *prescriptive*, *predictive*, *descriptive* or *explanatory*.

A model of an as yet untested construction can be used to predict how it will behave initially. Similarly, to establish what kind of properties a non-existing original will possess, reality can be *imitated* by using the model in a *simulation*. With regard to the time aspect, models may be either *static* or *dynamic*. Models which exclude the influence of time are typically static, while those including time are dynamic. In a *dynamic simulation* a model is rapidly exposed to a continuous series of inputs as it passes through an artificial space and time. Simulation is only possible if there exists a mathematical model, a virtual machine, representing the system being simulated. Today this machine is represented by the computer.

A special kind of simulation is *gaming* which most often involves decision making in critical situations. The decisions relating to hypothetical conditions are taken by real decision-makers. Sometimes the situation includes a *counter-measure team* which increases the degree of difficulty.

Widely-known laws, principles, theorems and hypotheses

Systems knowledge of a more general nature, particularly within systems behaviour, has been expressed in different laws, principles, theorems and hypotheses. This knowledge is considered to be within the core of General Systems Theory even if its origin is to be found in another area. Some of the formulations presented here cover a broad scope of systems aspects and are extensively applicable, although most of them concern living systems.

The different parts of General Systems Theory are reiterated below, beginning with the laws.

- **The second law of thermodynamics** In any closed system the amount of order can never increase, only decrease over time.

- **The complementary law** Any two different perspectives (or models) about a system will reveal truths regarding that system that are neither entirely independent nor entirely compatible (Weinberg 1975).

- **The law of requisite variety** Control can be obtained only if the variety of the controller is at least as great as the variety of the situation to be controlled (Ashby 1964).

- **The law of requisite hierarchy** The weaker and more uncertain the regulatory capability, the more hierarchy is needed in the organization of regulation and control to get the same result (Aulin & Ahmavaara 1979).

The following general principles are valid for all kinds of systems.

- **System holism principle** A system has holistic properties not manifested by any of its parts. The parts have properties not manifested by the system as a whole.

- **Suboptimalization principle** If each subsystem, regarded separately, is made to operate with maximum efficiency, the system as a whole will not operate with utmost efficiency.

- **Darkness principle** No system can be known completely.

- **Eighty-twenty principle** In any large, complex system, eighty per cent of the output will be produced by only twenty per cent of the system.

- **Hierarchy principle** Complex natural phenomena are organized in hierarchies wherein each level is made up of several integrated systems.

- **Redundancy of resources principle** Maintenance of stability under conditions of disturbance requires redundancy of critical resources.

- **Redundancy of potential command principle** In any complex decision network, the potential to act effectively is conferred by an adequate concatenation of information.

- **Relaxation time principle** System stability is possible only if the system's relaxation time is shorter than the mean time between disturbances.

- **Negative feedback causality principle** Given negative feedback, a system's equilibrium state is invariant over a wide range of initial conditions.

- **Positive feedback causality principle** Given positive feedback in a system, radically different end states are possible from the same initial conditions.

- **Homeostasis principle** A system survives only so long as all essential variables are maintained within their physiological limits.

- **Steady state principle** For a system to be in a state of equilibrium, all subsystems must be in equilibrium. All subsystems being in a state of equilibrium, the system must be in equilibrium.

- **Self-organizing systems principle** Complex systems organize themselves and their characteristic structural and behavioural patterns are mainly a result of interaction between the subsystems.

- **Basins of stability principle** Complex systems have basins of stability separated by thresholds of instability. A system dwelling on a ridge will suddenly return to the state in a basin.

- **Viability principle** Viability is a function of the proper balance between autonomy of subsystems and their integration within the whole system, or of the balance between stability and adaptation.

- **First cybernetic control principle** Successful implicit control must be a continuous and automatic comparison of behavioural characteristics against a standard. It must be followed by continuous and automatic feedback of corrective action.

- **Second cybernetic control principle** In implicit control, control is synonymous with communication.

- **Third cybernetic control principle** In implicit control, variables are brought back into control in the act of, and by the act of, going out of control.

- **The feedback principle** The result of behaviour is always scanned and its success or failure modifies future behaviour.

- **The maximum power principle** Those systems that survive in competition between alternative choices are those that develop more power inflow and use it to meet the needs of survival.

Living systems also follow a number of main systemic principles, foremost in connection with preserving stability. The twelve below have been defined by ***Watt*** and ***Craig*** in their book *Surprise, Ecological Stability Theory* (1988).

- **The omnivory principle** The greater the number of different resources and of pathways for their flow to the main system components, the less likely the system will become unstable. In other words: spread the risks or 'Don't put all your eggs in one basket.'

- **The high-flux principle** The higher the rate of the resource flux through the system, the more resources are available per time unit to help deal with the perturbation. Whether all resources are used efficiently may matter less than whether the right ones reach the system in time for it to be responsive.

- **The variety-adaptability principle** Systemic variety enhances stability by increasing adaptability.

- **The flatness principle** The wider their base in relation to their number of hierarchic levels, the more stable organizational pyramids will be. A larger number of independent actors increases stability.

- **The system separability principle** System stability increases as the mean strength of interaction between components is decreased. Stability is enhanced by separating the elements of the system from one another.

- **The redundancy principle** Generally, arithmetic increases in redundancy yield geometric increases in reliability. In self-organizing systems, negative feedback regulates reproduction where too little redundancy leads to the species dying out and too much to over-reproduction.

- **The buffering principle** Stability is enhanced by maintaining a surplus. An unused reserve cannot however help the system.

- **The environment-modification principle** To survive, systems have to choose between two main strategies. One is to adapt to the environment, the other is to change it. The beaver, for example, changes the environment for its own benefit.

- **The robustness principle** The ability of a system to passively withstand environmental change may derive from simple physical protection or it may involve a complex of mechanisms similar to those used by the butterfly to overwinter as a pupa.

- **The patchiness principle** The lack of capacity to use a variety of resources leads to instability (the external counterpart to the omnivory principle). Rule-bound systems, stipulating in advance the permissible and the impermissible, are likely to be less stable than those that develop pell-mell.

- **The over-specialization principle** Too much of a good thing may render systems unstable in the face of environmental change. It is through this principle that the conflict between the parts and the whole is played out.

- **The safe environment principle** Based upon the environment-modification principle, it states the importance of creating a permanently stable environment whereby the system is protected from change.

The following theorems together contribute further perspectives to the above material.

- **Gödel's incompleteness theorem** All consistent axiomatic foundations of number theory include undecidable propositions.

- **Redundancy-of-information theorem** Errors in information transmission can be prevented by increasing the redundancy in the messages.

- **Recursive-system theorem** In a recursive organizational structure, each viable system contains, and is contained in, a viable system. (In a military context, for example, it says that group, platoon, company, etc., all have the same functions and structure as integrated parts of the battalion.)

The following hypotheses have been selected from the approximate total of one hundred in Miller's book *Living Systems* (1976), in which he introduced the General Living Systems theory (GLS). Their social and managerial implications are obvious.

- A system's processes are affected more by its suprasystem than by its supra-suprasystem or above, and by its subsystems more than by its subsubsystems or below.
- The amount of information transmitted between points within a system is significantly larger than the amount transmitted across its boundary.
- The larger the percentage of all matter/energy input that a system consumes for the information processing necessary to control its various system processes, as opposed to matter-energy processing, the more likely the system is to survive.
- Strain, errors and distortions increase in a system, as the number of channels blocked for information transmission increases.
- In general, the farther components of a system are from one another and the longer the channels between them are, the slower is the rate of information flow among them.
- The higher the level of a system the more correct or adaptive its decisions.
- Under equal stress, functions developed later in the phylogenetic history of a given type of system break down before the more primitive functions do.
- The greater the resources available to a system, the less likely there is to be conflict among its subsystems or components.
- The vigour of the search for resolutions of conflicts increases as the available time for finding a solution decreases.
- A component will comply with a system's purposes and goals to the extent that those functions of the component directed towards the goal are rewarded and those directed away from it are punished.

Some of the axioms related to GST can be found on p. 31. The nature of axioms has been explicitly expressed in Gödel's theorem, presented on p. 64.

Some generic facts of systems behaviour

System behaviour, such as is expressed in the formulations on the foregoing pages, can always be related to the concept of complexity. The more complex a system, the more intricate its behaviour. It is, however, necessary to bear in mind that, given enough time and space, even the simplest system can produce quite unexpected and surprisingly complex phenomena. To emphasize the characteristics of a complex system, the following comparison between simple and complex systems has been made by ***R. Flood*** and ***M. Jackson*** (1991).

Simple systems are characterized by

- a small number of elements
- few interactions between the elements
- attributes of the elements are predetermined
- interaction between elements is highly organized
- well-defined laws govern behaviour
- the system does not evolve over time
- subsystems do not pursue their own goals
- the system is unaffected by behavioural influences
- the system is largely closed to the environment

Complex systems are characterized by

- a large number of elements
- many interactions between the elements
- attributes of the elements are not predetermined
- interaction between elements is loosely organized
- they are probabilistic in their behaviour
- the system evolves over time
- subsystems are purposeful and generate their own goals
- the system is subject to behavioural influences
- the system is largely open to the environment

A large system *per se* normally signifies a greater complexity inasmuch as more subsystems and more processes are in operation simultaneously. The degree of organization inherent in the system, defined as predetermined rules guiding the interaction, is another basic determinant. Nonlinear and stochastic processes with many feedback loops of higher order and time delays are also important.

A complex system often behaves in an unexpected manner and the relations between cause and effect are often difficult to understand. Measures taken to understand or control may sometimes yield the opposite of our intentions. Measures seemingly reasonable in the short term often prove to be harmful in the long run. Human interference with delicate regulation mechanisms may cause changes which lead quite abruptly into a new state, essentially irreversible and continuing for a very long time.

A system is generally less sensitive to external structural influences than to internal. The complex system can nonetheless be unsusceptible to changes in the internal parameters. Increases or decreases in their values are neutralized by many kinds of negative feedback and these changes have little influence on system behaviour. Linear negative-feedback systems are either stable or unstable regardless of the input signal applied. Nonlinear feedback systems may be stable for some inputs but unstable for others.

The general unifying forces keeping a systemic hierarchy together vary according to their evolutionary direction. Systems which evolve upward possess strong cohesive forces joining the subsystems and are thereby less easily disrupted. Lower levels generally function better than the later established suprasystem. In systems which evolve downward, developing specialized subunits, the suprasystem is stronger than the younger subsystems.

All feedback systems are apt to oscillate, affecting the behaviour, to a lesser or greater extent. Within the existing network of coupled variables each variable has a highest and lowest threshold. Within these limits the system can vary freely; if they are exceeded, disorder and finally collapse will occur. If they do not maintain a defined value, other variables will then occupy the available variation space and take over. For the systems researcher it is a constant problem to know whether certain feedback loops reveal significant differences, or merely amplify insignificant ones.

In Figure 2:19, A shows a state of stable oscillations where the input is a feedback from the output. This occurs when the feedback (thin line) has a phase which is the opposite to the system disturbance and is of equal amplitude. In B the oscillations are damped and diminish when the feedback is less than the output. Finally, a feedback signal inducing corrective action greater than the error will amplify the same, causing growing oscillations and instability, according to C in Figure 2:19.

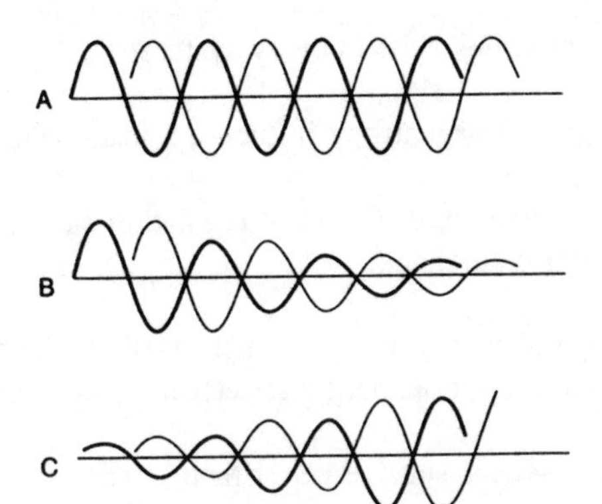

Figure 2:19 Different oscillation patterns in feedback systems (from Tustin 1955).

A special kind of system behaviour is associated with system growth and adaptation. The introduction of a unique input at some critical time may suddenly permit a semi-organized system to organize itself into a hierarchy and to grow. The general unifying forces keeping a systemic hierarchy together vary according to their evolutionary direction. From these facts follows the inevitable conflict in all growing systems, that existing between the supra- and subsystems.

The fact that living systems *per se* tend to complicate their interactions with their surroundings over time, is an inherent feature of their growth. This applies also to non-living systems – the growth of a computer network is a good example. Growth is mostly a consequence of an adaptation; bigger systems survive better than small ones. Inasmuch as it is not possible to adapt to everything, a system is prevented from growing to infinite size. Beyond a certain point integration and communication problems within the system exceed the benefits of large size.

Structural growth and the associated increase of size demand a specialization or modification of some components which in turn produce emergent systemic properties. A wider range of functions in a system with specialized subsystems makes it better equipped to cope with even unforeseen difficulties. The total system may thus have a longer life than its subsystems.

Review questions and problems

1. Systems theory tries to define a law about laws. Does such a law already exist within other disciplines?

2. Give some examples of dissimilar definitions of the concept system and point out some of the main differences.

3. What is the difference between an abstract system and a conceptual system?

4. Do emergent properties exist in every system?

5. How do entropy and negentropy influence a man-made system?

6. Describe how systems with feedback regulation can neutralize disturbing forces not known beforehand.

7. Einstein's proposition regarding mathematical models seems to have a certain connection to Heisenberg's uncertainty principle. Explain why.

8. Why do bigger systems survive better than smaller systems?

3 A Selection of Systems Theories

- **Boulding and the Hierarchy of Systems Complexity**
- **Miller and the General Living Systems theory**
- **Beer and the Viable System Model**
- **Lovelock and the Gaia Hypothesis**
- **Teilhard de Chardin and the Nōosphere**
- **Taylor and the Geopolitic Systems Model**
- **Klir and the Epistemological Hierarchy of Systems**
- **Laszlo and the Natural Systems**
- **Cook and the Quantal System**
- **Checkland and the Systems Typology**
- **Jordan and the Systems Taxonomy**
- **Salk and the Categories of Nature**
- **Powers and the Control Theory**
- **Namilov and the organismic view of science**

'It is the theory which decides what we can observe.' (*Albert Einstein*)

In Chapter 2 some of the properties peculiar to theories of general systems were identified. When considering the systems theories and models presented in this chaper we find that they fulfil the demands to varying degrees. A feature common to all (with one exception) is that they are hierarchies of both complexity and size. Another common property is that their structure seems to exist at all levels and on all scales.

It is obvious that the originators of these systematic structures shared interests, albeit from various perspectives. With their backgrounds as philosophers, sociologists, biologists, physicists, etc., they formulated their theories of systems in different terms. Their fundamental and also common belief is that certain main aspects of the world can be tied together through the myriad of systems into one rational scheme.

Nothing can be understood in isolation. All systems theories are therefore in a sense *explanatory structures* intended to correspond to something in the real world. As an explanatory structure certain theories give the framework for a specific systems methodology, described in Chapter 9. A further distinctive mark of systems theories is their focus on the principles of *organization per se*, regardless of what it is that is organized and the quality thereof.

A common practice when presenting systems theories is to begin with Boulding, Miller and Beer. Their concepts and vocabulary are easily comprehended and they provide a language with which to penetrate the other theories. This order of presentation can be seen as an attempt to lay a foundation with the most general and overarching theories on which the more specialized can be built.

Boulding and the Hierarchy of Systems Complexity

As one of the founding fathers of the Systems Movement, ***Kenneth Boulding*** presented his classical paper, *General Systems Theory – The Skeleton of Science*, in 1956. In this paper the author is deeply concerned about the existing over-specialization of science and the lack of communication between the different areas. He proposes as a way of overcoming this dilemma the use of an over-arching language of concepts, and an arrangement of theoretical systems and constructs in a hierarchy of complexity. This should be a 'system of systems' possible to use in all areas. Each scientific area studies some kind of system and a classification is necessary if a general methodology for their study is to be developed.

The first level in Boulding's hierarchy is the level of static structures and relationships or, using his term, **frameworks**. Examples are the arrangement of atoms in a crystal, the anatomy of genes, cells, plants or the organization of the astronomical universe. These can all be accurately described in terms of static relationship, of function or position. Organized theoretical knowledge in many fields emanates from static relationship, which is also a prerequisite for the understanding of systems behaviour.

The second level is called **clockworks**. The solar system offers an example of a simple dynamic system with predetermined motion. Machines such as car engines and dynamos, as well as the theoretical structures of physics, chemistry and economics, belong to this category; they all strive for some kind of equilibrium.

The third level is that of control mechanisms or **cybernetic systems**. A thermostat with its teleological (purpose geared) behaviour is an often used example for this level; another is the regulation called homeostasis. This level is characterized by feedback mechanisms with transmission and interpretation of information.

The fourth level is the level of the **cell** or the self-maintaining structure. Since life begins and develops here this is also called the open-system level. Life presupposes throughput of matter and energy and the ability to maintain and reproduce itself.

The fifth level can be called the **plant** level and is identified by its genetic/societal processes. The main qualities of these processes are differentiation and division of labour respectively, and mutual dependence between the various components for both. Since the life processes of the plant level take place without specialized sense organs, the reaction to changes in the environment is slow.

The sixth level is the level of the **animal** where the main characteristics are various degrees of consciousness, teleological behaviour and increased mobility. Here a wide range of specialized sensors convey a great amount of information via a nervous system to a brain where information can be stored and structured. Reactions to changes in the environment are more or less instantaneous.

The seventh level is the **human,** wherein the individual is defined as a system. Man possesses in addition to the qualities of the animal level, self-consciousness. This is a self-reflective quality: he not only knows but knows that he knows. A consequence of this is the awareness of one's own mortality. Another quality is a sophisticated-language capability and the use of internal symbols through which man accumulates knowledge, transmitting it from brain to brain and from generation to generation.

The eighth level is defined as that of **social organization**. A single human being, that is, one isolated from fellow human beings, is rare. The units of this level are the assumed *roles* and these are tied together by the channels of communication. Many cultural factors - value systems, symbolization through art and music, complex areas of emotion, *history* - are significant for this level.

The ninth level is the **transcendental**, or that of the unknowable. While one can only speculate about its structure and relationships, it is presupposed that this level does exhibit systemic structure and relationship.

A *system of systems* or a hierarchy of complexity indicating the relationships between the different levels is shown in Figure 3:1.

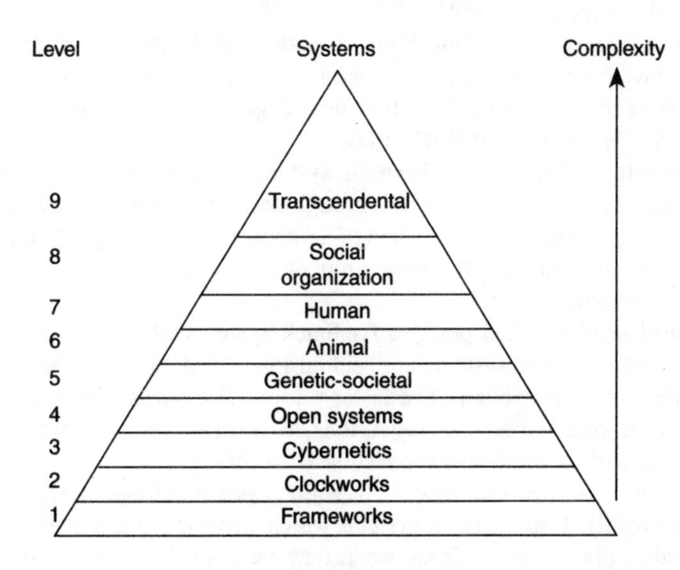

Figure 3:1 A hierarchy of systems complexity according to Boulding.

A closer examination shows that the first three levels belong to the category of physical and mechanical systems and are mainly the concern of physical scientists. Classical natural science is most at home at the clockwork level. Levels of the cell, the plant and the animal are typically the levels of biologists, botanists and zoologists. The next two levels of human and social organization are mainly the interest of the social scientists. Speculation concerning the nature of transcendental systems belongs primarily to the area of philosophy.

One of the motives behind Boulding's hierarchy was to present the status of the scientific knowledge of his time. Obviously, relevant theoretical models existed up to and including the cell level. Higher levels have only rudiments of relevant models. Boulding emphasized the gap between theoretical models and pertinent empirical referents which in his opinion were deficient within all levels. Another idea was that valuable knowledge can be obtained by applying low-level systems knowledge to high-level subject matter. This is possible inasmuch as each system level incorporates all levels below it.

In 1985 Boulding presented *The World as a Total System*, a book in which his original systems hierarchy is reworked and the levels are extended to eleven and are given slightly different names and contents. The first and most basic category in the level hierarchy can be described as a static and descriptive structure of space. The next category contains the description of dynamic systems – that is, of changes in static structures over time. The final category contains explanatory systems which, besides the pattern of structure in space and time, explain several basic regularities. Levels of complexity and regulation and control of the different subsystems are also to be found in the new hierarchy. All levels have their counterparts in the real world.

The first level consists of **mechanical systems** controlled by simple connections and few parameters. The connections are in mathematical terms seldom more complex than equations of the third degree. Examples are the laws of gravitation, Ohm's law and Boyle's law.

The second level is that of **cybernetic systems**. These more complex systems strive towards a state of equilibrium through negative feedback. Such a process exists in living bodies under the name of *homeostasis* and is dependent upon the processing of information. The basic units of *receptor*, *transmitter* and *effector* are always present.

The third level is called **positive feedback systems**. Owing to the nature of positive feedback these systems are seldom long-lived; they accelerate toward breakdown, or breakthrough. The faster a forest fire burns, the hotter it gets, and the more one learns, the easier it is to learn more. Evolution can also represent an anti-entropic positive feedback process.

The fourth level is the level of **creodic systems** (from Greek, meaning 'necessary path'). It includes all systems which strive towards a goal and which may be called planned in a wide sense as they are guided by some kind of initial plan. The morphogenesis in the development of both egg to chicken and the economy of a society can illustrate this level.

The fifth level is one of **reproductive systems** which implies that genetic instructions guide both reproduction and growth. Besides that for individuals, a reproductive process takes place in social organizations. Language and printed matter disseminate ideas; a member who is promoted, retired or dismissed is, through mechanisms for role occupancy, replaced.

The sixth level concerns **demographic systems** and consists of populations of reproductive systems. A population is a collection of comparable objects, not necessarily identical but similar enough to create a meaningful classification. A biological population increases through birth and decreases through death. If birth and death rates are equal and if normal age distribution is paralleled by the survival distribution, the population is said to be in equilibrium, a situation which is however rare.

The seventh level, **ecological systems,** consists of a number of interacting populations of different species. The size of a population is determined by its own structure and the size of its competitors. If in a given environment of other populations a specific population has reached equilibrium, it is said to occupy an *ecological niche.* If more populations are stable in their interactions they form an ecological system. The tropical rainforest is a useful example.

Ecological interaction between different populations takes place by means of *mutual competition*, *cooperation*, or *predation*. In a sense a similar interaction takes place between the artificial ecosystems of human artefacts, for example cars predate on people. The main difference is that the biological organisms can reproduce themselves, human artefacts cannot. Since a 'genetical structure' of artefacts exists in human minds as well as in artificial memories, artefacts can be said to reproduce in a figurative sense ('nōogenetics').

The eighth level is that of **evolutionary systems**. Such systems can be both ecological, changing under the influence of selection and mutation, and artificial, obeying the same influence but in the transferred sense of new ideas. The fact that the evolutionary process moves towards ever-increasing complexity may be seen in the emergence of human self-consciousness or in the growth of the city.

The ninth level, **human systems**, differs from other living systems owing to the superior information-processing capability of the brain. Advanced pattern recognition and communication abilities with speech, writing and the use of sophisticated artefacts are distinctive marks. (There are more species and sub-species of human artefacts than plants in Linné's sexual system.)

The tenth level is that of **social systems**, a result of the interaction between human beings and/or their artefacts. They arise thanks to the capability of human minds to form images and to convey complicated concepts from one mind to another. An interactive learning process where various types of experience and evaluations are communicated through the system is essential. The nature of these interactions can be classified as *threat*, *exchange* or *integration*. The social activity itself may be classified as belonging to economic, political, communicative and integrative systems. Processes of mutation and selection are at work both within the mass of human individuals as well as among their artefacts. The biological concept of an empty niche makes sense also when speaking of these artefacts. Cars fill up empty spaces all over the world; CocaCola competes successfully with a myriad of available beverages.

The nature of social systems and their internal and external interactions has been thoroughly dealt with in Boulding's book *Ecodynamics* (1978). In Figure 3:2 the resulting combinations of threat, exchange and integrative systems are represented.

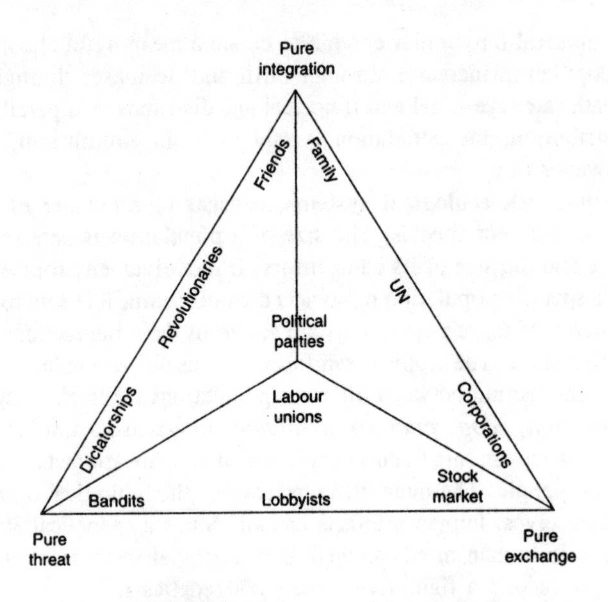

Figure 3:2 Combination of threat, exchange and integration in society. (From Boulding 1978)

The eleventh level is that of **transcendental systems**. Here certain religious or philosophical experiences may serve, at least in part, as examples. Being a level of the unknowable, the eleventh is one of speculation.

The hierarchical relationships between the different levels are shown below in Figure 3:3.

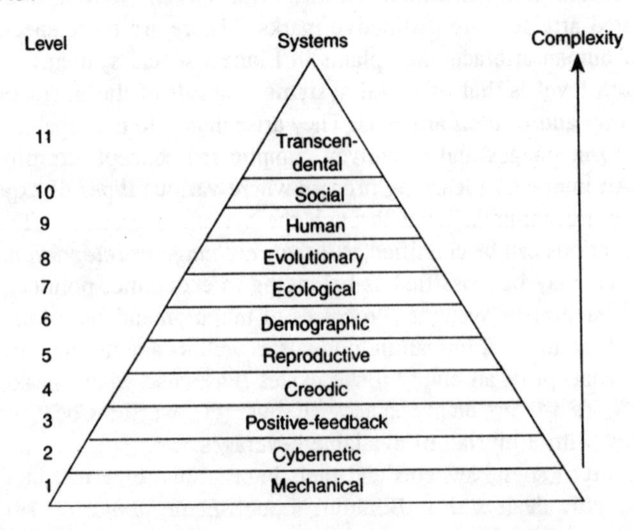

Figure 3:3 The second hierarchy of systems complexity according to Boulding.

Miller and the General Living Systems theory

James Miller, an American psychologist and psychiatrist, became one of the most prominent scientists within systems science after the publication of his book *Living Systems* in 1978. The General Living Systems theory or GLS theory presented in the book is concerned with a special subset of all systems, the living ones. It is general, or universal, in that it cuts across species, size of systems and type of behaviour and is interdisciplinary in that it integrates both biological and social science, ranging from cellular chemistry to international relations.

GLS theory recognizes the following five kingdoms:

- monerans
- protistans
- fungi
- plants
- animals

A living system maintains within its boundary a less probable thermodynamic energy process by interaction with its environment. Such a process is called **metabolism** and is possible through the continuous exchange of *matter* and *energy* across the system boundary. This process also gives the energy necessary for all essential activities, such as reproduction, production and repair. The metabolism or processing of *information* is of equal importance, making possible regulation and adjustment of both internal **stress** and external **strain**. See Figure 3:4.

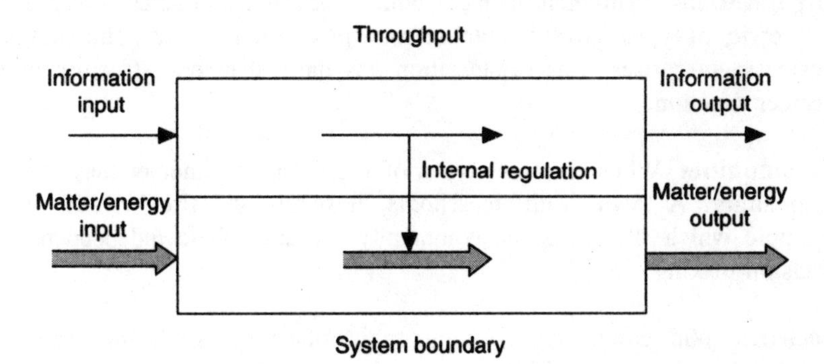

Figure 3:4 Throughput in a living system of matter/energy and information.

All living systems, irrespective of species, consist of remarkably similar organic molecules and a general evolutionary progression toward increasing complexity. Starting with the amoeba and finishing with the United Nations, living systems can be divided into eight very real and concrete hierarchical levels. Each new level is regarded as being higher than the preceding; it comprises all lower-level systems and is more differentiated. The vital system components of one level are systems in their own right on the level below. In other words, the larger and higher levels with their component lower-level subsystems constitute a *suprasystem*. Miller employs the metaphor of a ship's cable: a single unit which can be separated into the ropes of which it is twisted. These, in turn, can be unravelled revealing the finer strands, strings and threads.

Each level has its typical individual structure and processes. The levels are distinguished by the following labels:

- **Cells** These are composed of non-living molecules and multimolecular complexes and represent the least complex system that can support essential life processes. Cells exist either free living or as specialized components of the organs or tissues of organisms.

- **Organs** All organs are composed of structures of cell aggregates. An example is the liver.

- **Organisms** Their components are organs. This level includes multi-cellular plants and animals.

- **Groups** Two or more organisms which interact form a group. No higher level than this exists among animals. Structure and processes discernible among social insects such as bees and ants are more similar to those of the group than those of the next level.

- **Organizations** With their main components of groups, organizations present a diversity of types: governmental sectors, private universities, churches and business enterprises. The organization has more than one structure in its decider function.

- **Communities** When different types of organizations interact they form a community. A town, with its schools, hospitals and fire brigade, is an example which illustrates the community's characteristic independence in decision making.

- **Societies** With components of communities of various kinds and functions, this level is defined by Miller as *totipotential*. This indicates that, within itself, it contains all the essential capabilities as a self-subsistent system. A typical society is the nation which claims and defends a territory.

- **Supranational systems** Here two or more societies cooperate to a certain extent in decision making and in submitting power to a decider superordinate to their own. This level is represented by blocks, coalitions, alliances and pacts. NATO represents a single purpose; the European Union and the United

Nations exemplify a multipurpose supranational system. Societies express themselves through delegates within the decider function.

While living systems according to Miller's theory are all systems which support the phenomenon of life, social theory has no definite answer to whether social entities that are not organisms have a real existence 'in nature' or not. Some scientists prefer to see them as methodologically necessary theoretical constructs with no existence *per se*.

It has also been questioned if systems at levels above the organism can be considered to be alive: the components have no physical connections (as in the lower levels). The component individuals need physical contact similar to that of mechanical systems but only for sexual union or physical combat. Furthermore, at the higher levels, components can move from one system to another. These systems also include a great many nonliving components or *artefacts* that are crucial for the system. Non-living components in the form of *prostheses*, for example plastic aortas, can be present at lower levels, where even free-living components such as the white blood cells exist.

The importance of spatial cohesion is dependent upon the nature of the system. To preserve themselves as an effective existing unit the members of a riot squad quite naturally have to work shoulder to shoulder. The family as a system may function well even although grown-up children are geographically dispersed. The occupied space and boundaries of such systems are entirely conceptual, that is, they exist in the minds of people and not in physical reality.

Lack of physical cohesion among components of a living system is often compensated for by advanced communication systems which tie the components together. The low frequency acoustic long-distance communication between big whales can in this respect be compared with man's corresponding telecommunication system.

Living systems at all levels have some critical processes essential for living and reproduction. GLS theory identifies 20, each of which is performed by special units or components of systems. These exist at each of the eight levels, except for the two units necessary for learning, that is, associator and memory, which only exist in the animal organism. A critical process lacking in one system can sometimes be performed by some other. The 20 subsystems are divided into three groups: for the processing of matter/energy/information, of matter/energy and of information. (A system can survive without a reproducer, but not without any of the other subsystems.)

All living systems have to carry out the 20 essential subsystem functions in order to survive. Some systems, in which both structure and processes of some of the essential subsystems are missing, survive by substituting with either their own processes or processes of other systems at the same or different levels, or in Miller's words by *dispersing missing processes*. All the subsystems are presented below in the order and with the numbers given in 1990, when the timer was added (Miller 1990).

Subsystems processing matter/energy/information:

#1 **Reproducer** is capable of giving rise to other systems similar to the one it is in.

#2 **Boundary** at the perimeter of a system holds together the components which make up the system, protecting them from environmental stress, and which exclude or permit entry to various types of matter/energy and information.

Subsystems processing matter/energy:

#3 **Ingestor** brings matter/energy from the environment across the system boundary.

#4 **Distributor** carries inputs from outside of the system or outputs from own subsystems to each component within the system.

#5 **Converter** changes certain inputs to the system into forms more useful for the special processes of that particular system.

#6 **Producer** forms stable associations among matter/energy inputs to the system or outputs from its converter. The materials so synthesized are for growth, damage repair, or replacement of components of the system. They also provide energy for constituting the system's outputs of products or information markers or moving these to its suprasystem.

#7 **Storage** retains deposits of various types of matter/energy in the system, for different periods of time.

#8 **Extruder** transmits matter/energy in the form of products and waste out of the system.

#9 **Motor** moves the system or parts of it in relation to its whole or partial environment, moves components of the environment in relation to each other.

#10 **Supporter** maintains the proper spatial relationships among components.

Subsystems processing information:

#11 **Input transducer**, with its sensory function, brings markers bearing information into the system, changing them into other matter/energy forms suitable for internal transmission.

#12 **Internal transducer**, with its sensory function, receives, from all subsystems or components within the system, markers bearing information concerning significant alterations to the same. It changes these markers to other matter/energy forms which can be transmitted within the system.

#13 **Channel and net,** composed of a single route in physical space, or multiple interconnected routes, transmits markers bearing information to all parts of the system.

#14 **Timer** transmits to the decider information about time-related states of the environment or of components of the system. This information signals to the decider of the system or deciders of subsystems when to start, stop, alter the rate, or advance or delay the phase of one or more of the system's processes, thus coordinating them in time.

#15 **Decoder** alters the code of information input to it through the input transducer or the internal transducer into a 'private' code that can be used internally by the system.

#16 **Associator** carries out the first stage of the learning process, forming enduring associations among units of information in the system.

#17 **Memory** carries out the second stage of the learning process, storing various types of information in the system for different periods of time.

#18 **Decider** receives information inputs from all other subsystems and transmits to them information outputs that control the entire system. This executive decider has a hierarchical structure, the levels of which are called *echelons*.

#19 **Encoder** alters the code of information inputs from other information-processing subsystems, from a 'private' code used internally by the system into a 'public' code which can be interpreted by other systems in its environment.

#20 **Output transducer** puts out markers bearing information from the system, changing markers within the system into other matter/energy forms which can be transmitted over channels in the system's environment.

A main feature of the GLS theory is a level/subsystem table of living systems with 160 cells (8 levels × 20 subsystems). Here the components of the subsystems are listed for the various levels, altogether 153, with 7 missing as not recognized. The arrangement has a remarkable resemblance to the periodic table of the elements and in a sense it has a similar function for living systems. For an extract from the table, see Figure 3:5.

Subsystem Level	Reproducer	Boundary	Ingestor	Distributor	Converter	Producer
Cell	DNA and RNA molecules	*Matter-energy and information*: Outer membrane	Transport molecules	Endoplasmic reticulum	Enzyme in mitochon-drion	Chloroplast in green plant
Organ	Upwardly dispersed to organism	*Matter-energy and information*: Capsule or outer layer	Input artery	Intercellular fluid	Gastric mucosa cell	Islets of Langerhans of pancreas
Organism	Testes, ovaries, uterus, genitalia	*Matter-energy and inform-ation*: Skin or other outer covering	Mouth, nose, skin in some species	Vascular system of higher animals	Upper gastro-intestinal tract	Organs that synthesize materials for metabolism and repair
Group	Parents who create new family	*Matter-energy*: Inspect soldiers; *Information*: TV rules in family	Refreshment chairman of social club	Father who serves dinner	Work group member who cuts cloth	Family member who cooks
Organization	Chartering group	*Matter-energy*: Guards at entrance to plant; *Information*: Librarian	Receiving department	Assembly line	Operators of oil refinery	Factory production unit
Community	National legislature that grants state status to territory	*Matter-energy*: Agricultural inspection officers; *Information*: Movie censors	Airport authority of city	County school bus drivers	City stockyard organization	Bakery
Society	Constitutional convention that writes national constitution	*Matter-energy*: Customs service; *Information*: Security agency	Immigration service	Operators of national railroads	Nuclear industry	All farmers and factory workers of a country
Supra-national System	United Nations when it creates new supranational agency	*Matter-energy*: Troops at Berlin Wall; *Information*: NATO security personnel	Legislative body that admits nations	Personnel who operate supranational power grids	EURATOM, CERN, IAEA	World Health Organization

Figure 3:5 Extract from the level/subsystem table of living systems.
(From 'Introduction, The Nature of Living Systems', J. Miller, *Behavioral Science*, 35, 3, 1990).

To adapt to a continually changing environment and to handle stress from both within and without, living systems embody *adjustment processes*. System variables are hereby kept within their normal ranges and the system as a whole, as well as its subsystems, maintains homeostasis in spite of continuous changes. However, adjustment entails a cost (time, money, etc.). The least costly processes are engaged in the first instance, followed by the more resource demanding when necessary. The continuous use of adjustments which are more costly than necessary is a kind of *systems pathology*. Another kind of process, the *historical*, is the system life cycle. This includes growth, development, maturation, ageing and death.

Higher levels among living systems are both larger and more complex and have therefore structure and processes not existing at lower levels. This phenomenon is called *emergents* and gives these systems a better capability to withstand stress and adapt or exploit a greater range of environments.

The role of information processing has been thoroughly dealt with in GLS theory. Living systems, at least on higher levels, depend upon a full flow of the following three types of information in order to survive:

- information of the world outside,
- information from the past,
- information about self and own parts.

Living systems recognize three types of codes of increasing complexity used in the information metabolism.

- An **alpha** code is one in which the ensemble or markers is composed of different spatial patterns, each representing a coded message or signal. Signal agents like *pheromones* belong to this category.
- A **beta** code is based on variations in process such as temporal or amplitudinal change or a different pattern of intensity.
- A **gamma** code is used when symbolic information transmission takes place, as in linguistic communication.

Information processing involving these codes often gives rise to stress, known as *information input overload*, especially on higher system levels. This phenomenon is well-known in Western urban civilization which in a sense exists in this state continuously. Here the citizen withdraws within his own world as a result of adjustment processes or a quite logical attempt to survive in a seemingly chaotic world of excessive information. Of the possible adjustment processes, the following are the most common in connection with overload.

- **omission** neglecting to transmit certain randomly distributed signals in a message
- **error** incorrect transmission of certain parts of a message
- **queueing** delayed transmission of certain signals in a message
- **filtering** certain classes of messages given priority
- **abstracting** finer details in the message omitted
- **multiple channel transmission** parallel transmission over two or more channels
- **chunking** messages with intelligible meaning organized in chunks rather than individual symbols
- **escape** information input cut off

Information stress which occurs with *information input underload*, or sensory deprivation (for example, a patient in a respirator – the isolation syndrome), also requires adjustment. While these processes incur some costs, the system normally begins with the least costly. The possible adjustment processes in this case are the following.

- **sleepiness,** eventually falling asleep
- **inability to think clearly,** growing irritation, restlessness and hostility
- **daydreaming**, use of fantasy materials to supply information inputs
- **regression,** revert to childlike emotional behaviour
- **hallucinations,** sensation of 'otherness'
- **psychological breakdown,** total mental disorder

Some of the main features of GLS theory have now been outlined. This theory, itself too far-reaching to be presented in detail here, has many practical applications. For this purpose the chart of living systems symbols is used when depicting system flows between the twenty essential subsystems, as shown in Figure 3:6.

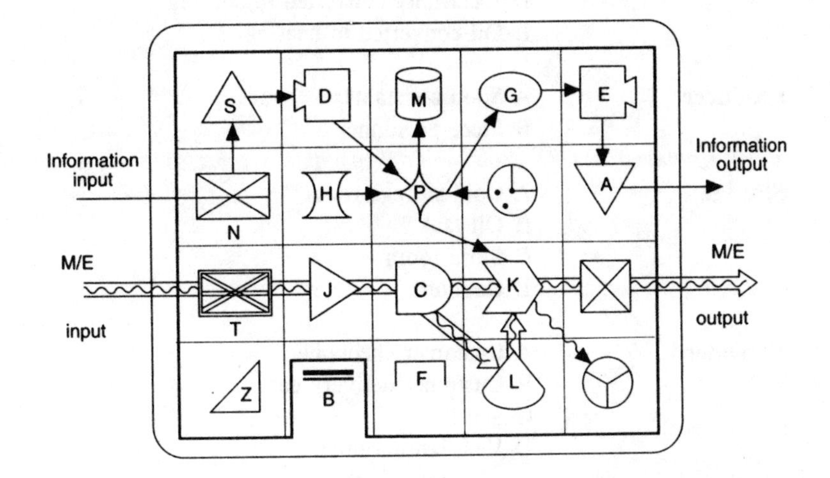

Figure 3:6 Flows between essential subsystems of a living system.

When applying GLS theory to a problem of the real world a key task is to identify the components of the 20 essential subsystems for the object, within its level. The example below shows how an identification can be established for a supermarket.

# 1 **Reproducer**	A The recruitment function B The owner's mental venture model C Accumulated funds
# 2 **Boundary**	A The walls of the building B Decrees and laws
# 3 **Ingestor**	A Oil inlet B Electricity inlet C Air inlet D Merchandise inlet E Commodity inlet F Water inlet
# 4 **Distributor**	A Corridors B Pipework C Cables
# 5 **Converter**	A Unpacking B Arrangement C Raw food converted to fast food

	D Electricity converted to cooling E Oil converted to heating
# 6 **Producer**	A Shop assistants B Store personnel
# 7 **Storage**	A Cold-storage room B Oil tank C Store room D Counters
# 8 **Extruder**	A Customer checkouts B Customer delivery car C Sewer D Ventilation outlet E Waste disposal
# 9 **Motor**	A Employees B Cleaning agency C Machinery
#10 **Supporter**	A Building site B The building C Artefacts D Caretaker agency
#11 **Input transducer**	A Assistant at counter B Office (mail, phone, papers) C Representative at sales meeting
#12 **Internal transducer**	A Weighing B Testing C Tasting D Customer conversation
#13 **Channel and net**	A Written internal messages B Internal telephone C Telephone D Internal conversation
#14 **Timer**	A Seasonal variations B Changes of customer's taste C Book-keeping and budget cycles D Business-hour regulations

#15 **Decoder**	A Owner B Clerks
#16 **Associator**	A Owner B Accountant
#17 **Memory**	A Archive B Register C Owner
#18 **Decider**	A Authorities B Trade union C Owner
#19 **Encoder**	A Marketing consultant
#20 **Output transducer**	A Information media

Beer and the Viable System Model

The Viable System Model (VSM) was first presented in 1972 by ***Stafford Beer*** in his book *Brain of the Firm*. The VSM, a model just as complicated as its prototype the human body, consists of elements analogous to extremities, backbone, nerves, nerve centres and brain. These counterparts are the five managerial subsystems. As a survival instrument of the organism, the human nervous system has to process an excess of information and regulate a tremendous number of variables. Taking its way of functioning as a starting point, Beer calls his model *neurocybernetic:* it is directed toward information flows and communication links within the enterprise. The way information circulates in the various channels gives a hint of how both the organization as a whole and its different parts perform in relation to their goals.

A *viable* system has the properties of self-repair, self-awareness, recursion and maintenance of identity. According to Beer the structure and working principles of a viable nervous system are applicable to all kinds of organizations for their regulation, adaptation, learning and development. When the performance is faulty, it is assumed that the cybernetic principles are being violated. Organizations exist in a very complex reality – 'a terrible mess' in the words of Beer. To handle their organization, managers have to tame the mess according to some basic principles presented in the model where the main concern is the control function and the concept of **variety**. The general solution to the problem of organizational complexity or variety is to use the fact that variety neutralizes variety. This is defined in the *Law of Requisite Variety* (see p. 61) thus: the variety of the control unit has to be at least the same as the variety of the governed system. A massive variety reduction is possible through *organizational recursion* which implies that every systemic level is a recursion (organizational copy) of its metasystem.

Variety has sometimes to be multiplied and sometimes to be damped, giving the terms **amplifier** and **attenuator**. A **transducer** is also necessary as a translator in a communication process which crosses several subsystem boundaries.

Using these key concepts, Beer formulated four principles which all viable organizations must obey.

- **The First Principle of Organization**
  Variety, diffusing through an institutional system, tends to equate; it should be *designed* to do so with minimum cost and damage to people.

- **The Second Principle of Organization**
  Channels carrying information between the management unit, the operation and the environment must each have a higher capacity than the generating subsystem.

- **The Third Principle of Organization**
  Whenever the information carried on a channel crosses a boundary, it undergoes transduction; the variety of the transducer must be at least equivalent to the variety of the channel.

- **The Fourth Principle of Organization**
  The operation of the first three principles must constantly recur through time, and without hiatus or lag.

All viable organizations also consist of the five subsystems discussed below. **System One** refers to those units that are to be controlled. These are exemplified by departments in a firm or subsidiaries in a group of companies. The basic organizational elements are shown in Figure 3:7.

The *square* encloses the managerial activity needed to run the organization. The *circle* encloses the operations that constitute the total viable system in focus. The *amoebic* shape represents the total environment. The *arrows* refer to the vital interactions between the three basic entities; each arrow stands for a multiplicity of channels whereby the entities affect each other. The amplifier is intended for the low-variety input and the attenuator the high-variety input, thereby balancing the variety.

The figure shows the dynamic content of any enterprise. Manipulation of the four Ms, *Men*, *Materials*, *Machinery* and *Money,* exists as part of the more fundamental management of complexity. Complexity and its measure, variety, indicate the number of possible states of a system. Variety in a complicated entrepreneurial system is calculated with the help of comparative statements (this has more variety than that) and of the arithmetic of ordinal numbers (this product is the second most profitable).

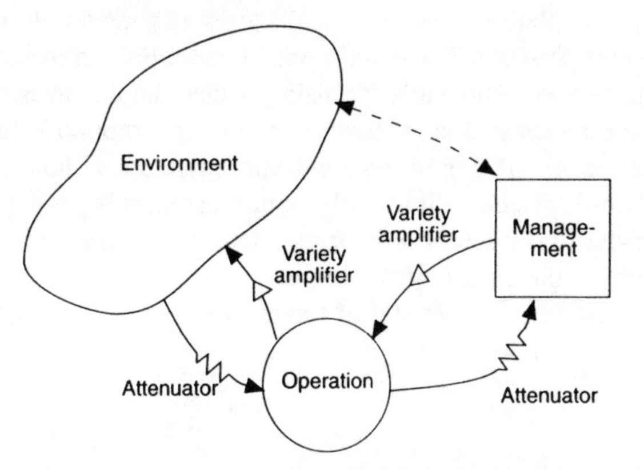

Figure 3:7 Basic organizational elements of the enterprise.

It is obvious that the square management box has a lower variety than the circle containing the operations, and that the circle in turn has lower variety than the environment. Variety is cut down, or attenuated, to the number of possible states that the receiving station can actually handle. Variety may also be increased, or amplified, to the number of possible states that the receiving station needs if it is to remain regulated and under control.

System Two coordinates the parts of System One in a harmonious manner. It comprises the information system necessary for decentralized decision making within System One and for problem solving between the separate System Ones. This is carried out through formal reporting and through people who build their own networks of contacts. Uncontrolled oscillations between the parts are dampened by System Two, a kind of service organ. A continuously working System Two is essential although its *requisite variety* only works in a dampening way. Audit is a typical System Two function creating a channel between System Three and the operational in System One.

System Three is the 'here and now' of the enterprise and its functional components are typically Marketing, Accounting, Personnel and so on. Two of its main tasks are maintaining the inner-connectivity of its own infrastructure and the exact configuration of System One. It also has to interpret policy decisions of higher management and to allocate resources to parts of System One. Its own policy *vis-à-vis* System One should be effectively implemented.

System Four is the forum of 'change and future'. While System Three handles the inside of the enterprise, System Four handles the outside, that is, the managing of external contacts, development work and corporate planning. The future does not happen, it has to be designed. This is the task of System Four.

The distribution of environmental information upward or downward according to its degree of importance is the responsibility of System Four. Urgent information and 'alarm signals' from the lower levels must be received and eventually forwarded to System Five.

System Five completes the system and closes the model. It monitors the operation balancing Systems Three and Four. System Five, a metalevel with an irregular appearance, is responsible for main policies and the investments for the infrastructure. Examples such as shareholders, the governors of a university, or the board of directors of a multinational enterprise show how System Five represents the 'whole system'. Since only significant signals passing through the filters of all of the lower levels will bring about a response, this level can be seen as representing the cortex of the brain.

A diagram of the complete model of the above is shown in Figure 3:8.

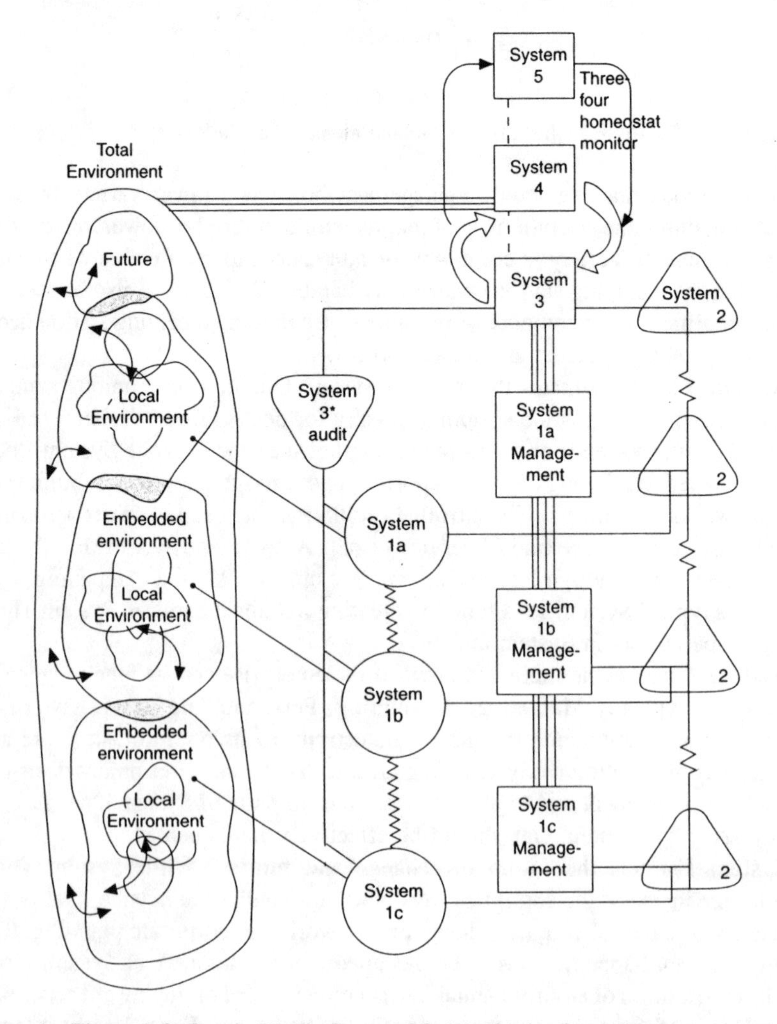

Figure 3:8 Beer's Viable System Model with its five subsystems.

For a full understanding of Figure 3:8 it is necessary to be acquainted with the **Recursive System Theorem**. In Beer's own words: *'any viable system contains, and is contained in, a viable system'*. To study this phenomenon is to consider a trio of viable systems: the organization we wish to study, the one within which it is contained, and the set of organizations contained within it (one level of recursion down). A most important feature of a viable system is this *self-reference,* as illustrated in Figure 3:9. Note that the connectivity between any recursive pair is the same. Any organization, already and quite properly depicted as belonging to a special level of recursion, also belongs to a number of other recursion levels.

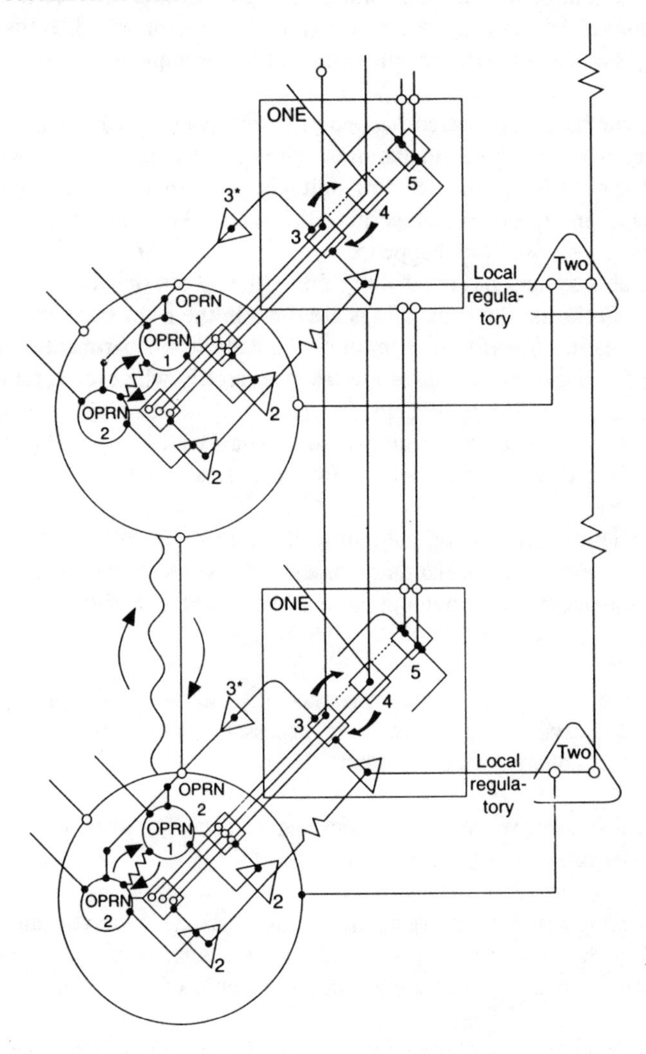

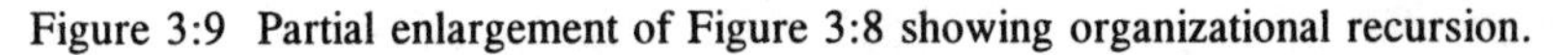
Figure 3:9 Partial enlargement of Figure 3:8 showing organizational recursion.

The ultimate aim of operation control is the maintaining of the homeostasis. A company's cost control, quality control, stock control and management inventory, among others, are examples of homeostatic regulation. This takes place through decisions in hierarchical order or a 'chain of command'. To emphasize the parallels between VSM and the human body let us examine the planning for future events in the enterprise. This is initiated by way of directions from a deciding organ (the brain or the annual meeting of shareholders). Information concerning decisions is transmitted to units (nerve-centre or the board of directors) which transform directions to execution orders (nerve impulses or messages). These orders are then interpreted by affected units (extremities or departments) which execute the order (the reaction or the directive). When everything is accomplished, the unit receives a task-completed message (response or report).

When something unexpected happens a reflex reaction takes over. This is a spontaneous response to a stimulus in the autonomous nerve system without the immediate knowledge of the deciding unit (the brain or the managing director). Afterwards an interpretation concerning the event takes place when the decider becomes aware of what had happened.

Every viable system has controlling units (the spleen or a chartered accountant) which check that the right things are done in the right manner. Some units have an evaluating function (the brain or the auditor's department). Inadequate results are fed back and measures are taken to rectify whatever is faulty.

The aim of the VSM is to demonstrate a well-functioning system. Several well-informed statements concerning organizational well-being, and malfunctioning, are delivered with the model. A few of these are as follows.

- Organizational freedom or autonomy is defined by means of interactions between operational horizontal forces and vertical unifying forces. If autonomy becomes synonymous with isolation the cohering forces will disappear.

- The degree of organizational coherence depends upon the purpose of the system. Metasystem intervention is necessary only to the degree which guarantees a coherent viable system.

- Complicated systems malfunction owing to inherent instability, not for a defined reason.

- Systems 2, 3, 4 or 5 of an organization often become autocratic and strive for viability in their own right, that is, they become bureaucratic. They should not be allowed to continue to function at the expense of the system as a whole.

- System 5 sometimes collapses into System 3 if System 4 is too weak.

- System 3 managers interfere too often in the management process of System 1.

- Internal communication channels and channels linking the organization to the environment are often underdimensioned for the flow necessary to ensure a viable system.

Lovelock and the Gaia Hypothesis

In 1972 ***James Lovelock*** and his co-worker ***Lynn Margulis*** introduced a hypothesis regarding the earth as a living superorganism. This hypothesis was later given the name of **Gaia** by the Nobel literature prize-winner ***William Golding***, who recognized the parallel between the mother earth of Greek mythology and Lovelock's idea. A further development of the hypothesis was presented in Lovelock's books *Gaia: A New Look at Life on Earth* (1979) and *The Ages of Gaia* (1988).

The hypothesis of Gaia is a contemporary expression of ancient wisdom concerning nature. The conception that the earth is 'living' is probably as old as humanity itself. Many cultures and religions, and speculators, since Aristotle had their convictions about the existence of a harmonious order in nature. The physician and alchemist ***Paracelsus*** (1493-1541) regarded nature as a complex organism in contact with an investigator. The Scot ***James Hutton*** (1726-97) wrote in 1788, in his book *Theory of the Earth,* about the restoring and healing forces of nature. In his lectures he is reported to have stated: 'I consider the earth to be a superorganism and that its proper study should be physiology.' In line with his predecessors, Lovelock has introduced the term *geophysiology.* This term stands for a systems approach to biospheric life analogous to the physiological study of an organism.

At the heart of the Gaia hypothesis is the fact that life creates its own milieu. Existing qualities of the atmosphere, oceans and continents are not prerequisites for life, they are instead consequences of life. Life influences the milieu in an extraordinary way, not possible for ordinary physical/chemical forces. Life has taken control of earth and transformed a lifeless planet into a self-regulating, self-sustaining organism. This organism continuously adapts the planet's physical, chemical and biological processes to maintain optimal conditions for the development of life. It does not strive to adapt to a changing environment.

The emergence of photosynthesis created an atmosphere in which countless new species could evolve. Photosynthesis itself could only provide enough energy for a vegetative existence, but with the emerging oxygen metabolism animals could generate the power they needed for movement and activity. The specific forms of life were all carriers of functions of the wholeness, following the principle that the most varied ecosystems are the most robust. Climate in turn has been dramatically affected by the presence of life.

Many ingredients vital to the function of the earth's atmosphere have been produced by the biosphere itself, such as oxygen which has a central position in the interpretation of the Gaia hypothesis. In one sense, the atmosphere is an artefact, like the honey or wax of a beehive. The processes of life cooperate to transform the planet into a safer place for life itself.

While other planets of our solar system are either extremely hot (Venus) or extremely cold (Mars), the earth has been successful in regulating its surface temperature for several billion years. It is estimated that the sun is 20 per cent hotter now than when life began 3.5 billion years ago. In spite of this increasing heat radiation, the surface temperature has been maintained within limits favourable for life.

Lovelock claims that the continued existence of life *per se* is therefore not a question of adjusting to a warmer environment; it is a question of maintaining the *status quo* through control of the environment. Symbiosis, the reciprocal actions between organisms and the environment, is seen as the source of evolutionary novelty and chief agent of natural selection. Even altruism or purposeful intent is expected to play a role. The Gaia phenomenon is a collective property of the growth, activities, and death of the innumerable populations that compose the biomass.

This contradicts the conventional evolutionary paradigm which asserts that life adjusts to the environment and sees the accumulation of chance mutations as the mainspring of development. The Gaia hypothesis in facts predicts that systems of this kind may arise *automatically* out of a mindless striving for survival. It adds however that life must be planetwide and powerful if it is to be present at all.

Lovelock states that biological control mechanisms always intervene when the earth grows too warm, too cold, too dry, too moist, etc. The various processes of life cooperate and regulate the global climate in order to create suitable conditions for life itself. Certain significant organisms can function on several levels and influence both the biomass, geology and chemistry of the earth. Their presence can alter land surfaces, water surfaces, albedo of clouds or act as nucleating agents for condensation. The chemistry of soil, water and air can be changed and thereby the distribution and transport of material may be changed. All parts of Gaia are interlocked through a series of complex feedback loops.

To illustrate the working principles of a global control mechanism Lovelock has constructed a computer model called Daisyworld. It is a hypothetical ecosystem consisting of black and white flowers. The black flowers thrive in a cold climate but, by acting as solar collectors, they induce heating. The white flowers thrive in a hot climate but reflect the sunshine and grow more numerous, thereby inducing a colder climate. Together these flowers can, without having an internal purpose, stabilize the temperature even if the incoming radiation increases.

According to the law of requisite variety (see page 61), to maintain stability a system must have at its disposal a sufficient number of regulatory mechanisms.

The following are some examples of such regulatory and protecting **global homeostasis mechanisms,** where life serves as an active control system.

- When solar radiation increases the oceans grow hotter. The water volume expands and its level rises. Nutritious substances from coastal areas are released into the water. With the consequent increase of plankton, dimethylsulfide accumulates in the oceans and evaporates into the atmosphere. Through this process the density of the heat-reflecting cloud cover is controlled and the temperature is held within a range of +5° to +35° C, the optimum for life. At the same time the growing amount of plankton absorbs the atmospheric carbon dioxide, decreasing the greenhouse effect.

- Seasonal variations of the atmospheric carbon-dioxide content are dependent upon the foliage of plants. In the summer the green foliage absorbs carbon dioxide; in the winter it is emitted by the decomposition process on the ground. This is a predominant climate regulator in the northern hemisphere with its greater land mass.

- An ozone layer in the upper atmosphere protects all kinds of life from devastating ultraviolet radiation from the sun. This layer has possibly been created by oxygen originating mainly from algae, themselves sheltered from radiation by the ocean water.

- Tropical rainforests control the circulation of water. The vast forests of the Amazon account for more than half of their own water circulation, that is, precipitation, outflow and evaporation. Devastation of these forests would result in huge amounts of water leaving the region and a global climate change. The rain produced by the forest is in itself vital; normally containing small amounts of ammonia it supplies acid to the soil for optimal vegetation.

- Lichen colonize inhospitable mountain and cliff surfaces and assist in their decomposition. Decomposition products form topsoil, a prerequisite for the life of other plants. The composition of the atmosphere is also influenced by this process.

- The maintenance of an average salinity of 3.4 per cent is essential for life in ocean water. While a concentration of 4 per cent would have produced quite different life forms, 6 per cent would make life in water impossible. Salts released by weathering should long ago have made the water too salty for life. It is assumed that excess salt is neutralized by the corals that build reefs, in turn forming lagoons wherein large quantities of salts are trapped as the seawater evaporates. Salt, together with other minerals, also provide raw materials for sheaths where bacteria live.

- The sudden drop in both carbon dioxide and temperature a billion years ago chilled the upper mantle. This in turn might have destabilized the lower crust causing the continental drift. Ongoing life processes may also drive geological plate tectonics necessary for the renewal of the earth surface.

- Major grass and forest fires may act as regulators, keeping the oxygen concentration in the atmosphere at an average of 21 per cent. Above an upper threshold of 25 per cent fires would destroy the biosphere.

The Gaia hypothesis postulates that the earth is a living organism - an enormous, complex and self-regulating web of life capable of influencing and regulating her environment. All living matter on earth, from virus to whale, from alga to tree, is regarded as constituting a single living entity. Gaia cannot be separated from the different parts of her body. The circulation of water from sea to land to sea is her flow of blood; her atmosphere is both the cuticle regulating temperature and moisture within her body and a protection against dangerous cosmic radiation from without. This tight coupling between life and environment and the constant preparedness for change and to adapt suggests distinctly different types of interactions, qualifying Gaia as a planetary super-organism.

Like other organisms Gaia has her life-cycle. Childhood implies a series of major changes alternating with long periods of relative little transformation. In adulthood the presence of life becomes the dominant quality of the planet. The niches are filled and the resistance to change is at its maximum. Inevitable old age is reached probably with the loss of internal heating and atmosphere. The parent star, the sun, reaches its Red Giant phase and Gaia is consumed by the expanding shell.

Lovelock's train of thought can be further clarified if Gaia is seen as a ninth level of the GLS theory, albeit of a slightly different logical type (Gaia existed before the emergence of the cell and is in a sense a zero level). The 20 essential subsystems of Gaia according to Miller's GLS theory can be identified as follows.

1 **Reproducer** A No evidence or
B Interplanetary travel?

2 **Boundary** A The earth's crust downwards
B The atmosphere upwards

3 **Ingestor** A The atmosphere from above
B Volcanoes from below

# 4 **Distributor**	A Winds B Streams C Grazing animals
# 5 **Converter**	A Mosses B Lichen C Plants D Animals
# 6 **Producer**	A Coral reefs B Volcanoes producing lava
# 7 **Storage**	A Soil (dead animals and plants) B Water C Oil D Minerals
# 8 **Extruder**	A Atmospheric gas discharge B Oceanic sedimentation
# 9 **Motor**	A Tidal water B Plate tectonics C Air streams
#10 **Supporter**	A The earth crust
#11 **Input transducer**	A Plants and animals (reacting to a daily rhythm, seasons, earthquakes, etc.)
#12 **Internal transducer**	A Plants and animals (reacting to climatic changes, pollution, etc.)
#13 **Channel and net**	A Grazing animal B Spread of plants
#14 **Timer**	A Movement of earth B Phases of the moon
#15 **Decoder**	A Plants and animals (reacting to the behaviour of other living beings)
#16 **Associator**	A Plants and animals (changed behaviour and functions)

#17 **Memory**	A Information stored in genes B Information stored in landscapes
#18 **Decider**	A Possibly the human race
#19 **Encoder**	A Atmospherical changes
#20 **Output transducer**	A Upper atmosphere (gas-release/radiation) B Changes in planetary albedo

With a growing population, human activities within the global ecosystem are ever increasing, converting natural ecosystems to self life-supporting systems. What some scientists call the 'The Greenhouse Civilization' has tampered with the planetary cycles of energy and materials. Agriculture, forest logging, industrialization and urbanization are devastating land uses which bring about extensive pollution of land, air and water. By converting natural negative feedback to unnatural positive feedback, human civilization has interfered with and blocked some of Gaia's control systems.

Gaia has thus been deprived of vitally important control systems and the preferences of the Greenhouse Civilization do not coincide with Gaia's. For example, Gaia's preferred temperature and ours are not the same. Interglacial periods like the present one, although existing for ten thousand years, may prove to be a global fever and the ice age may be the more stable state. The greenhouse effect itself is sometimes compared with a global fever used by Gaia to drive out devastating parasites.

Destructive activities of humans have also caused great damage to biodiversity. Loss of biodiversity by an accelerating extinction of all kinds of species is a devastating threat to the robustness of all the regulatory systems of Gaia. Furthermore, it will fundamentally and irreparably restrict our total understanding of Gaia as our planetary home.

The role of human beings in the Gaia system is, however, controversial among geophysiologists. Lovelock himself has a very cool attitude to humanity and its significance for Gaia; he views human civilization as largely irrelevant. We are neither owner nor preserver of the planet but are solely one species among others. Nature survives without man - man on the other hand cannot survive without nature. As a dynamic system, Gaia will always regain her balance but with no special preference for any particular life form. In comparison with earlier global catastrophes the Greenhouse Civilization leaves only a small scar in Gaia's skin. Ninety-nine per cent of all life forms that have existed since the beginning of life no longer exist; if we choose to break the rules of life, Gaia will exterminate us. In other words, humanity will not destroy Gaia, Gaia will destroy humanity. Gaia, as any living organism, cannot remain passive in the presence of threats to her existence and she will always be the stronger part.

Other scientists, standing outside of mainstream science, assert that Gaia maintains optimal conditions for life according to a purpose and thereby displays some kind of intelligence. Although life is fundamentally self-organizing and self-determined we do not know its final goal. We, as components of the superorganism Gaia, cannot (from without) study her ultimate goal (the cell sees a very limited part of the inside of the body!). Gaia has nevertheless acquired a goal with the emergence of the human mind; through human beings a central nervous system developed giving knowledge about herself and the rest of the universe. The capability to anticipate and neutralize threats to life itself has thereby been considerably augmented.

Seen through the eyes of ***Teilhard de Chardin*** (see p. 98) the ultimate goal of Gaia should be the creation of a nóospere, Gaia's thinking layer and the equivalent of the human neocortex. A human brain has at least 10^{12} neurons. Zoologists are of the opinion that a critical mass in brain volume is achieved at 10^{10} neurons, a level occurring in higher primates. If therefore each human being is regarded as a neuron in the brain of Gaia, and given that the world population will soon reach 10^{10}, a new kind of global consciousness might emerge. This new mental quality would not be the property of particular individuals; it would be manifested on a global level.

The interconnections between the neurons create the global nervous system, which has an analogue in the global communication and information networks. The slower hormonal communication system existing in all human bodies also has its counterpart: the worldwide post-office system. Even the two brain hemispheres, the right and the left, may have an analogue in the Western and Eastern spheres of culture.

Other scientists state that mankind can be likened to a malign cancer in Gaia's body. We have cut down the forests and polluted both water and air. But, like cancer cells, we are not aware that our own destructive activities will destroy Gaia's web of life – the basis of our own existence.

Which of the above views is the most realistic is at the moment not certain. It is quite possible that both views will be correct; nothing prevents a growing organism from struggling with cancer. Geophysiology is a new mode of synthesis. Time will tell....

Teilhard de Chardin and the Nōosphere

Already in the 1920s the impact of the growing state of knowledge and consciousness in humanity on the planet was the subject of an advanced discussion. Participators were among others the Russian geochemist Vladimir Vernadsky and the paleontologist and Jesuit Father ***Teilhard de Chardin*** (1881-1955). Together they coined the term **nōosphere** (from the Greek word *noos* for mind), implying an emerging mental sphere of intelligence covering the whole earth. This sphere, superposed on the biosphere, the sphere of life, is the main topic in Teilhard de Chardin's book *The Phenomenon of Man*. Both the nōosphere and the 'principle of Omega', also presented in the book, are examples of *finalistic* theories postulating some kind of cosmic teleology, purpose, or programme.

The book discusses the development of man as part of a universal process and presents both scientific, philosophical and theological perspectives. It is an attempt to synthesize the physical world with the world of mind, the past with the future and variety with unity. It has its starting point in the concept of *convergence,* denoting the evolutionary tendency of mankind towards non-specialization and unification. Man is the only successful race remaining as a single interbreeding species, that is, without splitting into a number of biologically separated branches. He has reached maturity; his body no longer changes. Transformation does take place but in mental and social contexts. Convergence is clearly visible in human cultures where differentiation is more and more levelled out, especially in the modern technological society.

The main force behind various modes of human convergence is the earth's shape as a spherical restricted surface. A rapidly growing number of individuals must share less and less available land. In spreading out around a sphere, man sooner or later meets his own kind; idea meeting idea produces an interconnected web of thoughts, the *nōosphere*. Human existence is thereby under the influence of a general *complexification* as ever more psychosocial energy is created. The impact of complexification is seen in nature, which before the appearance of man was only an unorganized pattern of ecological interaction. With mankind the mental properties of organisms become the most important characteristics of life.

As a step in the development of man, the present self-consciousness stage will transcend into a new mode of thought for his evolutionary future and the emergent nōosphere will grow ever stronger. This stage will integrate the self with the outer world of nature and also facilitate a complete exchange with the rest of the universe. Man has then reached his ultimate goal, or the *Point of Omeg*a, the final convergence. Teilhard de Chardin uses the metaphor of the meridians; as they approach the pole, science, philosophy and theology are forced to converge in one final point. The Point of Omega is here the opposite of a state of *Alpha* representing the elementary particles and their pertinent energies.

The origin of man and his development toward the Point of Omega is shown in Figure 3:10. The numbers on the left indicate thousands of years. The zone of convergence is not to scale.

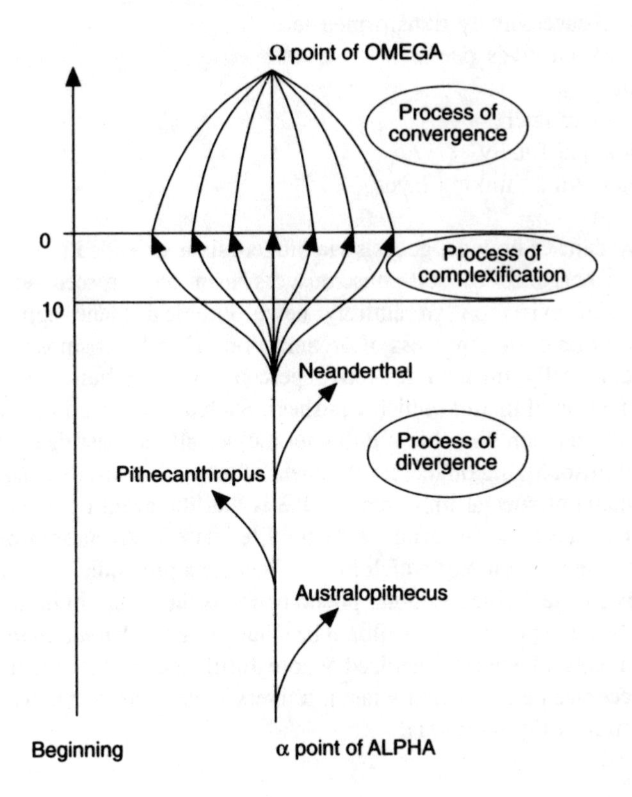

Figure 3:10 Human development toward the *Point of Omega* (from Teilhard de Chardin 1947).

To clarify his thoughts, Teilhard de Chardin, being a paleontologist, uses a geophysical model in which the nōosphere is the outermost of the six layers of *geogenesis*.

- barysphere
- lithosphere
- hydrosphere
- biosphere
- atmosphere
- nōosphere

These layers are a consequence of a goal-directed evolution where matter becomes conscious of itself in a self-organizing *biogenesis* by way of the following steps:

1 **energy** is successively transformed into
2 **matter** which gives rise to
3 **life** with
4 **instincts** and later
5 **thoughts** and finally
6 **nōosphere**, the thinking layer.

Here we can see how geogenesis via biogenesis is extended to psychogenesis. With the emergence of self-consciousness man has crossed some kind of threshold and exists on an entirely new biological plane approaching the culmination of a cosmic process of organization. Thereby cosmos fulfils its own goal systematically through reflective perception when building itself in the inverse direction of matter (which vanishes). Such a concentration of a conscious universe would be unthinkable if it did not include all the existing consciousness.

In his philosophizing on the development of man, Teilhard de Chardin focuses on phenomena of special importance. One is that the event of science provided the means to meet the material needs for life. Its effectiveness brought us the freedom of the modern city which in turn became a prerequisite for man, a predominantly mental being. Another phenomenon is the nature of human synthesis which leads to the nōosphere. Teilhard de Chardin states that *union differentiates* as the elements of every organized whole fulfil and perfect themselves. The element becomes personal first when it universalizes itself; a synthesis is not the disintegration of the individual.

Taylor and the Geopolitic Systems Model

Geopolitics as a discipline tries to explain the influence of geographical factors behind sociocultural organization and politics. Many models have been conceptualized within the area. One of the least specialized and easiest to understand was introduced by ***Alastair Taylor***, a professor of geography and political science at Queen's University in Canada (Taylor 1973).

The model has two parts: the levels or organization and the regulating mechanisms. It was designed mainly to explain the quantum leaps of social development in human history by correlating specific levels of societal and political organization with different stages of environmental control.

The first part of the model includes the following phenomena.

- the systemic levels of sociocultural organization

- processes demonstrating self-stabilization within a given level of organization and integration

- systemic transformation resulting in a sociocultural quantum leap across an environmental frontier

Five levels of organization, defined by the main man/environment relationships, are shown in Figure 3:11.

Interpreted horizontally, each column denotes exchange between the physical environmental factors and the control of the same through technology, science, transportation, communication and government.

Vertically, the five main levels of environmental control are presented as a geometrical sequence: *point, line, plane, volume*, as the control capabilities grow. All these levels are integrative or recursive, that is, they build upon each previous lower level. As time passes an increased complexity and heterogeneity takes place. Interpreted as a matrix, the model provides a time/space grid, giving societal/environment **quantization** when examined vertically and **stabilization** when examined horizontally.

The quantization takes place at a very slow rate on the lowest, food-gathering level and gains speed as passes through the different levels.

The stabilization shifts from **reactive–adaptive** to **active–manipulative** as it progresses through the different levels. Negative feedback ensuring overall stability is created by the social institutions and predominant general morality. Science and technology have a central role in creating the mechanism of positive feedback.

The second part of the model shows the importance of both positive and negative feedback processes and the interplay between them on all levels of sociocultural organization. The following phenomena are shown in Figure 3:12.

- biospheric and sociocultural inputs from the total environment

- the functioning of the existing system as a converter with numerous subsystems

- the regulation of the system's social and material output through negative and positive feedback

SOCIETAL LEVEL	SYSTEM OF ENVIRONMENTAL CONTROL		EMERGENT QUALITIES					
	EXPLETED SPACE	IMPLETED SPACE	PROPERTIES	TECHNOLOGY	SCIENCE	TRANSPORTATION	COMMUNICATIONS	GOVERNMENT
S_5	Three-dimensional (extra-terestrial)	Megalopis ('Ecumenopolis')	BELOW +	Electrical-nuclear energy Automation Cybernetics	Einsteinian relativity Quantum mechanics Systems theory	Supra-surface Inner space systems Outer space explorations Surface systems Sub-surface vehicles	Electronic transmission (simultaneously) throughout expleted space	'Ecumenocracy' (Supra-national policies) Multi-level transaction Sovereignty invested in global mankind
S_4	Two-dimensional (oceans, continents)	City	BELOW +	Transformation of energy (steam) Machine technology Mass production	'Greek miracle' Scientific method Newtonian world-view	Maritime technology and navigation Thalassic and oceanic networks Highway networks Railroad technology	Mechanical transmission (printing) Alphabet	National state system Emergence of democracy Sovereignty of state (as primary actor)
S_3	One-dimensional (riverine societies)	Town	BELOW +	Non-biological prime movers (wind, water) Metal tools Continuous rotary motion (wheel) Irrigation technics	Mathematics Astronomy	Sailboats Riverine transport Wheeled vehicles Intra- and inter-urban roads	Writing	Ancient bureaucratic empires Theocratic politics Sovereignty of god-kings
S_2	Particulated Universal (sedentary)	Village	BELOW +	Animal energy Domestication of plants and animals Polished stone tools Spinning Pottery	Neolithic proto-science	Animal transport Paths, village routes Neolithic seafaring	Ideograms	Biological-territorial nexus Tribal level of organization and decision-making
S_1	Undifferentiated Universal (Nomadic)	Cave/tent (intraterrestrial)		Human energy Control of fire Stone and bone tools Partial rotary action		Human transport Sleds Dug-outs, canoes	Pictograms	Biological nexus (family, hunting band, clan)

Figure 3:11 Levels of human societal organization.
(From Ervin Laszlo's *The World System*, copyright © 1972 by Ervin Laszlo. Reprinted by permission of George Braziller, Inc.)

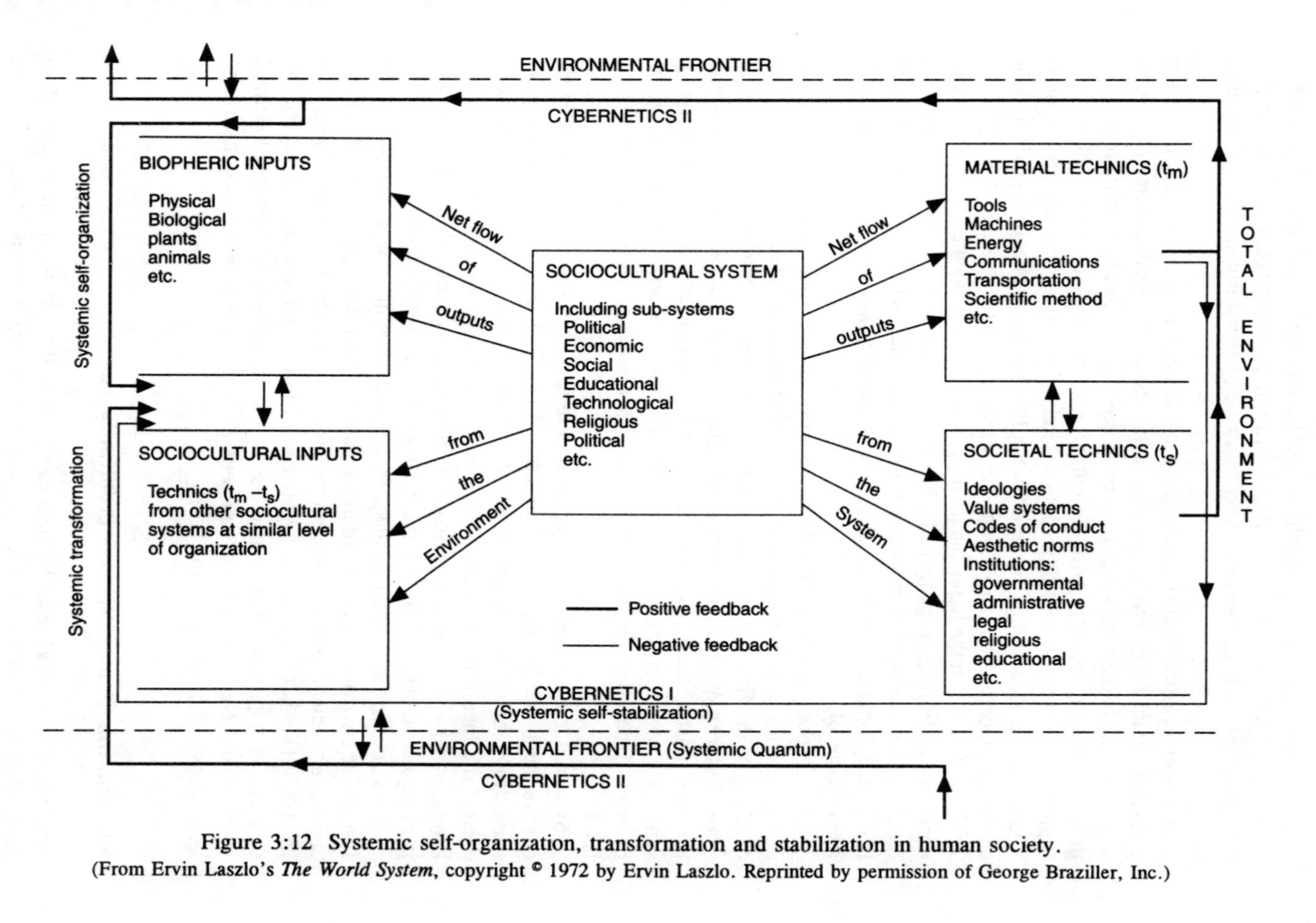

Figure 3:12 Systemic self-organization, transformation and stabilization in human society.
(From Ervin Laszlo's *The World System*, copyright © 1972 by Ervin Laszlo. Reprinted by permission of George Braziller, Inc.)

From the model it is possible to see how material and societal technique interaction results in systemic self-stabilization or transformation. In the model self-stabilization is denoted as Cybernetics I and transformation as Cybernetics II.

A major systemic transformation always implies an increase in information gathering, converted into knowledge and raising the total amount of system negentropy. Thereby the system's control capability expands, making it possible to cross existing borders with its output and rise to a new level of societal organization. Taylor has defined this kind of positive feedback as follows: 'Quantization occurs when deviation is amplified to the point where no deviation-correcting mechanism can prevent the rupturing of the basic systemic framework, that is, when the latter can no longer contain and channel the energies and thrust which have been generated.' Examples of self-stabilization with negative feedback can be detected in mature sub-hominid societies where the struggle for life is fully operative and the prevailing technique exercises maximal environmental control.

Transformation with positive feedback works in two ways: as self-organization and development and as systemic change with environmental quantization. An example of the first can be Eskimo societies showing an extensive command of available technique and environmental potential in the far North. By exploiting fire and usable tools like microliths they entered into a vital low-energy symbiosis with their extreme environment. Limits for such an existence given by way of negative feedback soon established societal stabilization on the S1 level.

Systemic transformation through access to less sophisticated tools in another environment can be seen in Mesolithic Asia. The use of stone sickles for harvesting, the domestication of wheat and barley and the taming of livestock created the 'neolithic revolution' on level S2. Here arose a new overall control, habitat patterns and social organization, with in turn a better division of labour, increased food supply, that is, the basis for a larger population.

A more dramatic quantum leap in connection with a new technique and favourable environment can be found in ancient China and Egypt. A food surplus and better societal organization opened the way for the 'urban revolution' on level S3. This is characterized by the rise of towns, refined administration, the division of population into classes, etc. These earliest civilizations, localized along rivers, could control and cultivate a whole valley and develop hydraulic technology. The city state was born.

Klir and the General Systems Problem Solver

According to ***George Klir*** his taxonomy of systems is intended to provide a pragmatically founded classification useful within various disciplines and engineering modes. It was first presented in his book *Architecture of Systems*

Problem Solving in 1985. The empirical application demonstrated in this work, the General Systems Problem Solver, GSPS, is a computer-based expert system.

Klir's conceptual framework builds upon strictly mathematical definitions (not presented here) and demands some experience in formal thinking to be wholly understood. Before defining a system as such, he identifies the system traits. These are compiled from the inherent variables (behaviour, states, etc.) and thereafter classified and formalized. If any of the primary traits participating in the definition change, the system also changes. The pertinent hierarchy consists of at least five fundamental levels of systems based on the following essential perspectives: that of an *investigator* and his environment, an investigated *object* and its environment, and an *interaction* between the investigator and his object. Each level embodies and supplements all lower levels.

Level zero, also called *source system* is determined by the way an investigator interacts with the investigated object. This interaction is partially guided by the preferences of the investigator but should include the following aspects: definition of a set of variables, a set of potential states for each variable and a description of the meaning of the variables with their states (real-world attributes and manifestations). The variables must be partitioned as either *basic* or *supporting*.

Basic variables can be either input or output variables and belong to *directed systems*. Systems without basic variables are called *neutral systems*. All supporting variables together form a *support set* wherein changes in the state of the basic variables take place. Examples of often-used supporting variables are space, time and aggregations of entities of the same kind (products, peoples, etc.). As the name zero level implies, it is a source of empirical data concerning the attributes of the object in focus.

Level one supplements the source system of level zero with data regarding the variables which are therefore called *data systems*. This level comprises all knowledge contained in level zero plus additional knowledge now available. Data is gained from measurement or observations or from the definition of desirable states.

Level two (and all higher levels) possesses knowledge of *support-invariant relational characteristics* of inherent data-generating variables for boundary conditions. Invariance implies that however the support set may change, certain features defining the function of the data set remain invariant to such changes. The support-invariant relation describes a process generating states of the basic variables of the support set. The level is thus one of *generative systems*.

Level three belongs to *structure systems* and is defined in terms of a set of generative systems, also called *subsystems* of the total system. The subsystems share different variables and interact in various ways.

Level four is described as *metasystems* and consists of a set of systems defined at levels 1, 2 or 3 and a support-invariant metacharacterization. Lower level systems must share the same source system.

Level five defines *meta-metasystems*. Here the metacharacterization is allowed to change within the support set according to support-invariant representation. *Metasystems of higher order* are defined in the same way.

A simplified overview of this epistemological hierarchy is shown in Figure 3:13.

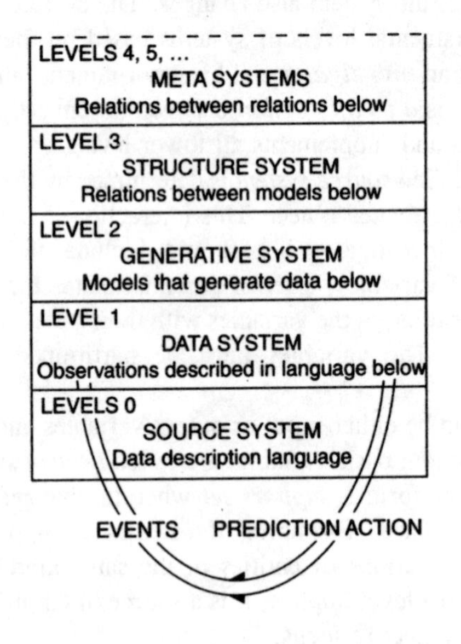

Figure 3:13 Klir's epistemological systems hierarchy.
(Reprinted with permission from G. Klir, *Architecture of Systems Problem Solving*, Plenum Publishing Corp., New York, 1985.)

Klir states that a system, a requirement or a problem is *identifiable* if it can be formulated in terms of his systems hierarchy of GSPS language. It is *admissible* if both identifiable and able to be handled in terms of a certain GSPS application. It is *solvable* if admissible and can be solved using the methodological tools available in a GSPS implementation.

The computer system built on Klir's conceptual framework enables the user to deal with systems problems. Figure 3:14 shows its four functional units: a control unit, a metamethodological support unit, a knowledge base and a set of methodological tools.

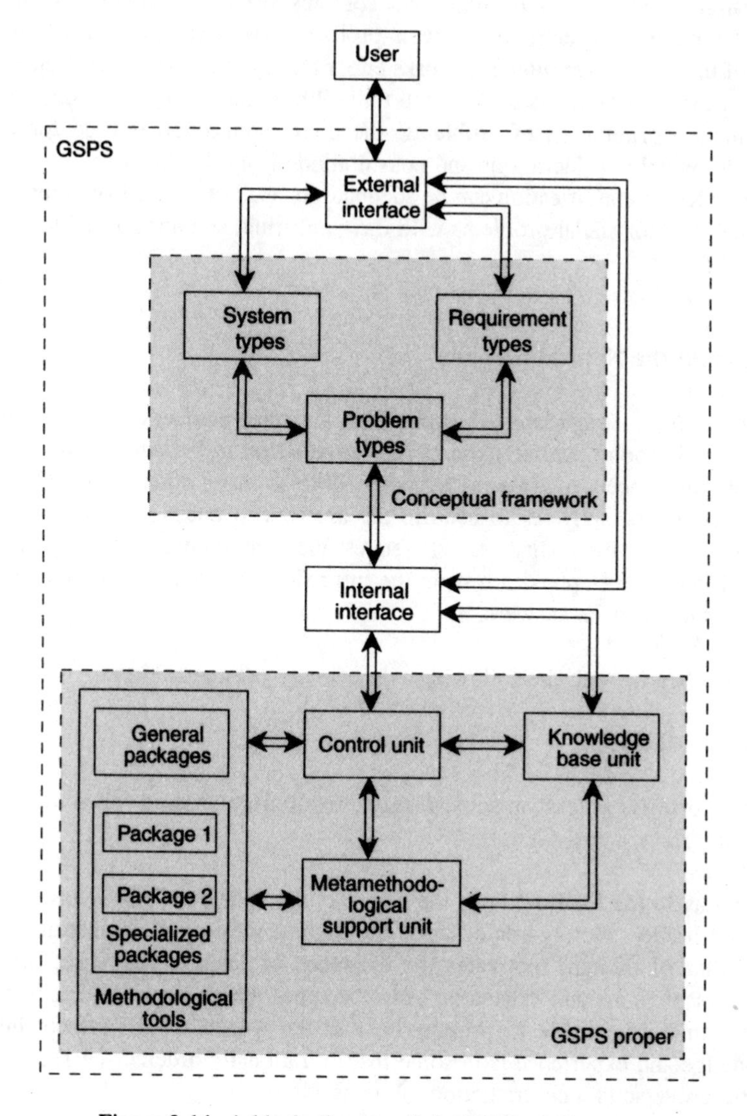

Figure 3:14 A block diagram of the GSPS architecture.
(Reprinted with permission from G. Klir, *Architecture of Systems Problem Solving*, Plenum Publishing Corp., New York, 1985.)

Methodological tools are methods having programs designed to solve admissible problems. Certain procedures specify the order in which individual algorithms are used. The *metamethodological support unit* contains information indicating how to order the problems according to how general they are. Information concerning problems which cannot be solved by the system is stored in the *knowledge base unit*. This unit also contains useful information regarding different kinds of system and systems problems, as well as relevant laws and rules of thumb. A *user interface* works either through the conceptual framework or via a direct connection to the main unit. While the conceptual framework is used in the formulation of problems, the direct connection is used for meta-methodological considerations and consultations from the database.

While Klir's classification can be adequate for well-structured situations, it is less adequate for social problems with their indistinct and unmeasurable nature.

Laszlo and the Natural Systems

Ervin Laszlo, a Hungarian philosopher, has written a number of books covering systems philosophy. In one from 1972, *Introduction to Systems Philosophy*, he presents his concept of **Natural Systems**. Like so many other philosophers, he turns first to metaphysics to acquire the answers to questions concerning the ultimate nature of reality. Laszlo states that, in contrast to religion, the propositions of metaphysics rest on the intrinsic coherence and simplicity - on the *elegance* - of its answers.

Laszlo begins with the following two primary presuppositions.

- The world exists.
- The world is, at least in some respect, intelligibly ordered (open to rational inquiry).

For Laszlo (as for Boulding, see page 31) the concept of order has its own, intrinsic beauty. He regards order as the highest ideal of the human mind and thus order of thought motivates the existence of science. Likewise, order in feeling inspires art and existential order becomes the mainspring of religion. It is also quite reasonable to presuppose that the world beyond present human knowledge and experience is in some respect rationally ordered. (A theory of a chaotic universe is a contradiction of terms.)

Laszlo continues with two secondary propositions.

- The world is intelligibly ordered in special domains.
- The world is intelligibly ordered as a whole.

In this world physical phenomena are viewed as systems according to modern mechanics or to field theory with complex subsidiary patterns such as the subsystems reflected in chemistry. Further counterparts of this view can be found in biology, where organisms are wholes forming a continuum demarcated by relative boundaries from still larger systems such as continental ecologies and social systems. From this starting point, Laszlo arranges all of the systems in two different planes: a macro-hierarchy and a micro-hierarchy.

In the *macro-hierarchy,* where the gravitational forces have the evolutionary role, the following entities of astronomy are found.

- planets and their sub-bodies
- stars
- star clusters
- galaxies
- galaxy clusters

In the *micro-hierarchy,* where the electrical and related forces are instrumental, entities of physics, chemistry, biology, ecology, sociology and internal relations are found.

- atoms
- molecules
- molecular compounds
- crystals
- cells
- multicellular organisms
- communities of organisms

While only a rudimentary state of knowledge exists of the macro-hierarchy, scientific knowledge of micro-evolution is extensive. Why the universe is fragmented into planets, stars and galaxies is not yet wholly understood. We have neither knowledge of the exact number of levels in the observable universe nor rational evidence that the series is infinite.

The theoretical hierarchical organization of nature with the micro-hierarchy superimposed on the macro-hierarchy is shown in Figure 3:15. The intersection between the two hierarchies is on the level of the atom. This integrated hierarchy with its different levels is called the *Natural Systems* by Laszlo.

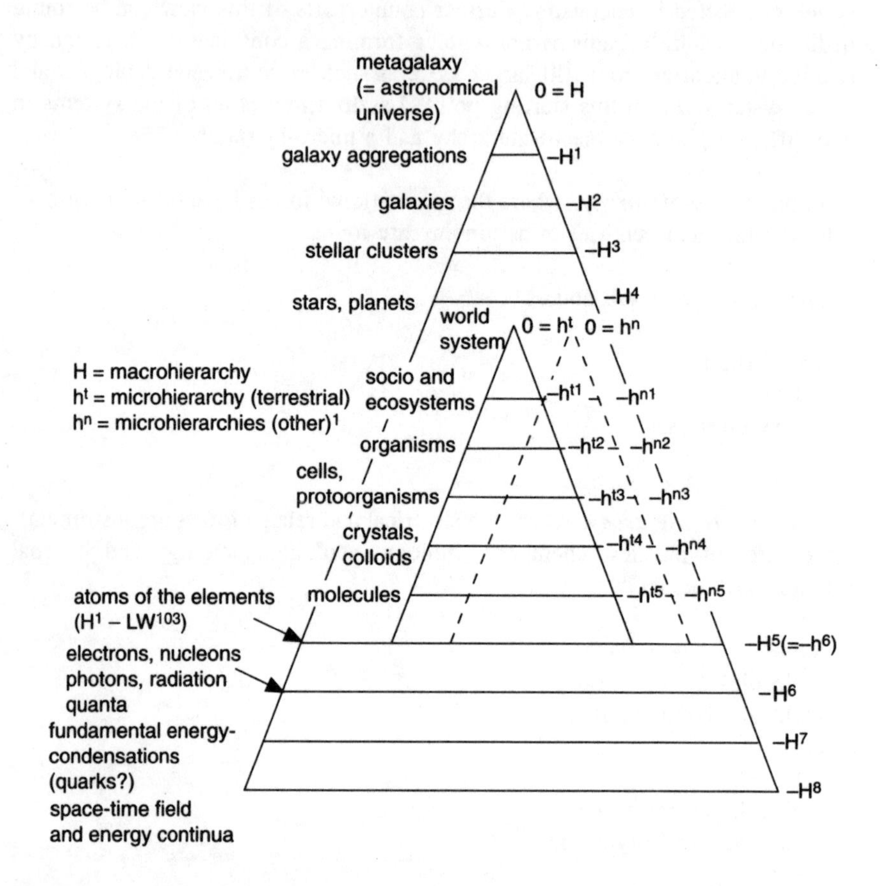

Figure 3:15 The Natural Systems hierarchy.
(Reprinted with permission from E. Laszlo, *Introduction to Systems Philosophy*, Gordon & Breach Publishing Group, Lausanne, 1972).

The development of natural systems from simple levels to more complex takes place according to adaptive self-organization, inevitably leading toward the known biological and psychological systems. Self-organization with its emerging complexification brings a decreasing stability to the system. Sudden disorganization is thus more probable at higher system levels than at lower ones. The inverse relation between structural complexity and self-stability is shown in Figure 3:16.

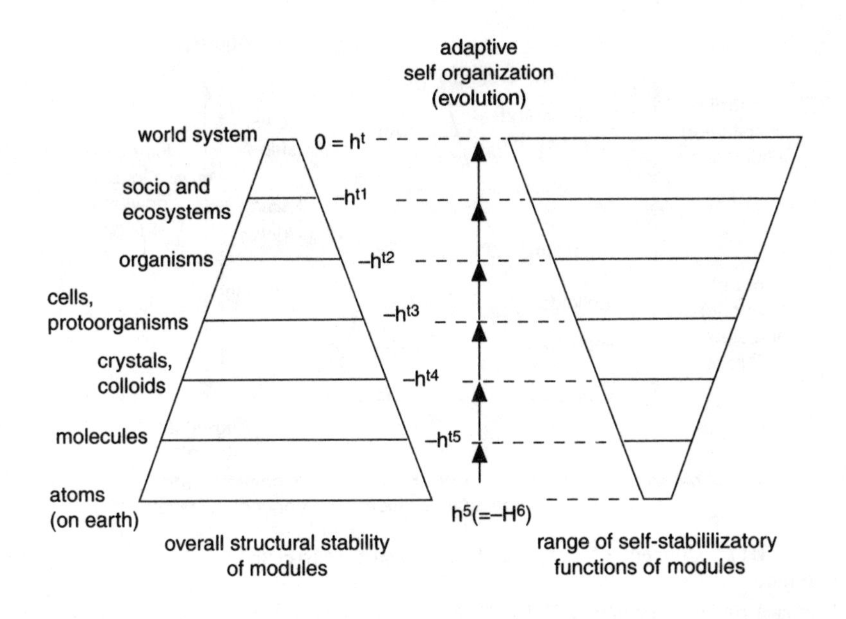

Figure 3:16 Relationship between structural complexity and self-stability in the micro-hierarchy.

(Reprinted with permission from E. Laszlo, *Introduction to Systems Philosophy*, Gordon & Breach Publishing Group, Lausanne, 1972).

The higher we climb the hierarchical ladder, the more diverse the functions and properties, albeit in a decreasing number of systems. Atoms exist in greater numbers than molecules but have fewer structural variations and fewer properties than the latter. Although organisms are fewer in number than molecules they exhibit a hugely greater range of functions and properties. Laszlo points to the about ten million existing species of plants and animals as an example of possible variation. While existing ecologies and societies are fewer in number than organisms, their properties and variations far exceed those of the single, or small groups of, organisms.

When examining the natural systems we find that they consist of both *things* and *relationships*. In Figure 3:17 the concept of things and relationships is related to the levels of micro-hierarchy. Things below but close to the human level are easy to grasp mentally; above this level the concept of things tends to transcend into relationships. On much lower levels of the hierarchy things are known only as an aggregate of smaller entities, such as crystals made up of molecules. Relationships become gradually weaker as we rise above the human level. From a personal point of view, the relationship *vis-à-vis* humanity is naturally more diffuse than one with a neighbour.

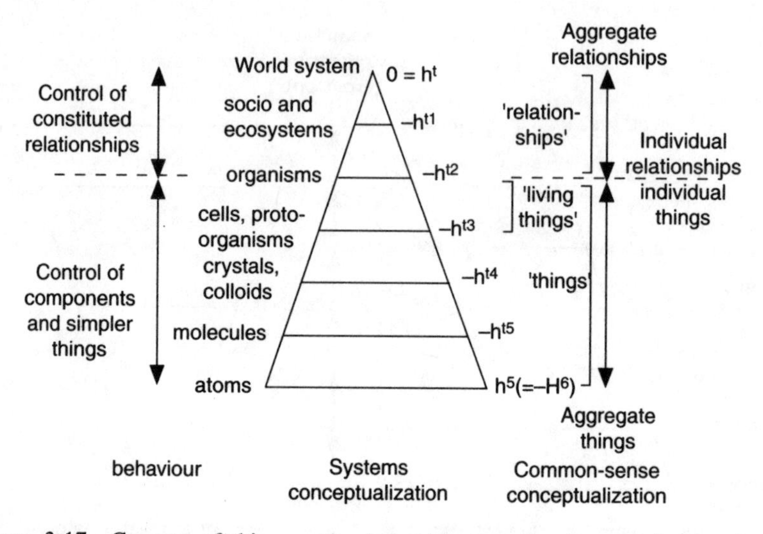

Figure 3:17 Concept of things and relationships correlated to the levels of micro-hierarchy.
(Reprinted with permission from E. Laszlo, *Introduction to Systems Philosophy*, Gordon & Breach Publishing Group, Lausanne, 1972).

Cook and the Quantal System

Norman Cook, a researcher in the area of neurosurgery, presented in 1980 his version of the Natural Systems in the book *Stability and Flexibility*. His theory describes how the functional duality of flexibility/stability influences five main levels of natural organization in our surrounding world. Together these levels are called the *quantal* by Cook.

These five levels represent the following fundamental units of nature and society and show structural and functional similarities.

- atom
- cell
- human organism
- human family
- nation state

Cook sees the atom as the only complete stable organization of space/time at the temperatures and pressures found in our own solar system. It is also the smallest entity known to physical science which deserves the name *system*. The cell is seen as the optimal unit of complex molecular organization on earth. It has emergent properties necessary to be the basic building block of life. Man, at the third level, is seen as the quintessence of biological evolution. The family system is seen as an entity unified by genetic factors as well as by the psychological need for self-expression. The nation state, finally, is seen as the natural result of interaction between families, clans, etc., fulfilling the human need for an arena where political and philosophical ideas can be realized.

The fundamentally antagonistic tendencies toward *constancy* and *change* on each level must be handled in such a way as to preserve internal stability of existing information. This also applies to flexibility and the possibility to share information with the external world. The functional duality of control allows for

- long-term viability of each system, a requisite for
- evolution of growing complexity within each system, making possible
- the functioning of the system within its suprasystem.

The control centres managing this functional stability/flexibility duality on the five levels are as follows.

- in the atomic nucleus, neutrons and protons
- in the cellular nucleus, DNA and RNA
- in the brain, the right hemisphere and the left hemisphere
- in the family the parents, woman and man
- in the government, the legislative branch and the executive branch

In his examination of the quantal, Cook finds a significant structural similarity between the atom and the other levels of his system. The neutron is specialized for atomic stability while the proton is specialized for contact with an external electron structure. No doubt a kind of genuine information exchange takes place in the atom allowing for a kind of systemic synergism.

As the building block of life, the cell is a highly ordered and complex system of biomolecular information. Information is stored in DNA molecules and expressed in the form of protein through RNA molecules. Genetic change is seen as a result of a creative cellular response to the biochemical milieu. Biological evolution *per se* may be seen as a continuous struggle to cope with the challenge of a changing and dangerous chemical environment. Cellular changes may then correlate with different environmental changes.

Man has a single personality in spite of his bicameral brain with two control elements. The coordination of both hemispheres of the brain is fundamental for the normal functioning of the unified individual. The right hemisphere is conventionally defined as soft, female, emotional; the left as dominant, male, intellectual. A right-sided life-style could be represented by Eastern meditational religions and use of drugs, music and poetry, and ecological awareness. The left-sided life-style could then be represented by the hard-working business man with his fixation on rational thinking, money and power.

The family is the natural means for one of the most essential activities of life: an expression of the human self through the distribution of one's own genes and distribution of one's modes of thought. To produce offspring who may (or may not) share one's qualities and world-view is to live beyond one's own biological existence. On the broader family level, anthropology has not found a culture where the social roles of woman and man have not been differentiated. There are also very few cultural exceptions to the dominant pattern of the internal, domestic female and the mobile, external man. As part of this stability/flexibility duality, the woman has been the idealist and the keeper of a stable, self-consistent world view. Man has been the practical half and the realizer of family values in the external world.

In the social system of the nation state Cook associates stability/flexibility to the metaphor of conservatism and liberalism. He also uses the analogy of the capitalist and the communist world to demonstrate some real consequences of the concept. The challenge on this level is to maintain a balance between *collectivism* and *individualism*. Most Western societies have approached this task by way of a division of the government into a legislative and an executive branch. The laws should be formed and enacted so as to reflect the needs of the nation rather than that of selected interest groups. Direct contact is seen as a threat to the autonomy of the legislative branch. The executive branch, however, should be in close contact with the needs of the people and facilitate processes such as lobbying, polling and voting. Stability on the national level is mainly concerned with protection of the ecological base to ensure the material existence of the nation and the existing rights of the people. Flexibility is a response to the basic needs of the people and a rational reaction to new economic and technological challenges.

A recapitulation of the control-centre duality and the resultant emergent qualities within the quantal levels is presented in Table 3:1.

Table 3:1 Control centre duality with emergent qualities in the quantal levels

System	Stability	Flexibility	Emergent quality
Atom	Protons	Neutrons	Atomicity
Cell	RNA	DNA	Life
Human organism	Left hemisphere	Right hemisphere	Mind
Human family	Woman	Man	Civilization
Nation	Executive branch	Legislative branch	Justice

Table 3:2 shows the malfunctions which arise when the predominance of one of the control-centre functions leads to an imbalance in the system.

Table 3:2 Control-centre predominance and associated systemic imbalances

System	Flexibility over Stability	Stability over Flexibility
Atom	Nuclear radioactivity (alpha decay)	Nuclear radioactivity (beta decay)
Cell	Malignant growth (RNA virus induced)	Benign growth (DNA virus induced)
Human organism	Schizophrenia	Manic depression
Human family	Male chauvinism	Feminism
Nation	Dictatorship	Bureaucratism

In the development toward higher levels of the quantal, Cook identifies the following four evolutionary stages.

- **The Primitive stage** has only one control centre and consequently neither functional duality nor emergent qualities representative of higher levels.

- **The Classical stage** has an emerging control duality but an insufficient internal communication.

- **The Modern stage** has an existing control duality and sufficient internal communication. The nature of its flexibility is however not correlated to the system as part of the surrounding metasystem.

- **The Completed stage** has full internal communication. The changes in the flexibility element are synchronized to changes in the system's environment.

The development of the four stages is of special interest with regard to the human organism. Cook's theory is supported by a similar train of thought originating from ***J. Jaynes*** in his book *The Origin of Consciousness in the Breakdown of the Bicameral Mind* (see bibliography). Here the correlation between human evolution and the cerebral hemisphere function is emphasized. With the liberation of the left hemisphere from total domination by the right we see the beginning of modern man with analytic, and later scientific, capability. An overview of this four-stage development within all quantals is presented in Table 3:3.

Table 3:3 The four developmental stages of the quantal

System	Primitive	Classical	Modern	Completed
Atom	Hydrogen	Unknown	Unknown	All elements
Cell	Viruses	Bacterial cells	Plant cells	Veterbrate cells
Human organism	Infrahuman	Proconscious	Conscious	Enlightened
Human society	Tribal hierarchy	Religious society	Secular society	Genuine democracy

Inasmuch as systems in the earlier stages of their development lack complete control duality, the suprasystem must answer for this function as part of a more complex, higher-level system.

Checkland and the Systems Typology

Peter Checkland, a British professor in systems science, published his book *Systems Thinking Systems Practice* in 1981. With this book he joins the scientists who have launched natural systems hierarchies, in his case called the Systems Typology. According to Checkland, the absolute minimum number of systems classes necessary to describe the existing reality is four. They are *natural*, *human activity*, *designed physical*, *designed abstract,* systems. (The typological map itself belongs to designed abstract systems.) Natural systems provide the possibility to investigate, describe and learn; human activity systems can be engineered and the designed systems can be created and employed.

Starting with **natural systems**, Checkland claims that 'they are systems which could not be other than they are, given a universe whose patterns and laws are not erratic'. Their origin is the origin of the universe and the processes of evolution. Within the natural systems there exists an obvious hierarchy from atoms to molecules. Combinations of molecules then give rise to a branched hierarchy.

Subatomic systems
Atomic systems
Molecular systems

give rise to

Non-living Systems	and	Living Systems
- inorganic crystals		- single cell creatures
- rocks		- plants
- minerals		- animals
		- ecologies

The next main level to be considered is that of **human activity systems** which have a tendency to integrate in such a way that they can be viewed as a whole. Most often other (designed) systems are coupled to them. An example is the oil industry with its oil rigs, tankers and fine-meshed distribution system. Out of the enormous number of human activity systems a few of the most typical can be noted.

- agricultural
- defence
- trading
- transportation

However, most fundamental on this level is the social system, represented by family, tribe, clan, etc. Typical here is the basic need of the members for mutual support within the frame of a community. In a sense, with their central structure, social systems belong to both natural and human activity systems. They have therefore been placed on the boundary between the two categories in the typology map of systems in Figure 3:18.

Designed physical systems can be defined as systems fitted with purpose of mind because a need for them in some human activity has been identified. To this category belong:

- individual tools
- individual machines
- other designed and fabricated material entities.

Designed abstract systems are various types of theological, philosophical or knowledge systems. While designed physical systems in principle can be produced by animals and insects (the bird's nest, the spider's web, the beaver's dam), designed abstract systems are only associated with human beings.

In Figure 3:18, all systems are related to each other on a global map. The difference in logic types between natural and human activity systems gives rise to separate kinds of investigations. The classical method of science with its observer standing outside is quite relevant for natural systems. When it comes to human activity systems, Checkland emphasizes the importance of the point of view influencing the observations.

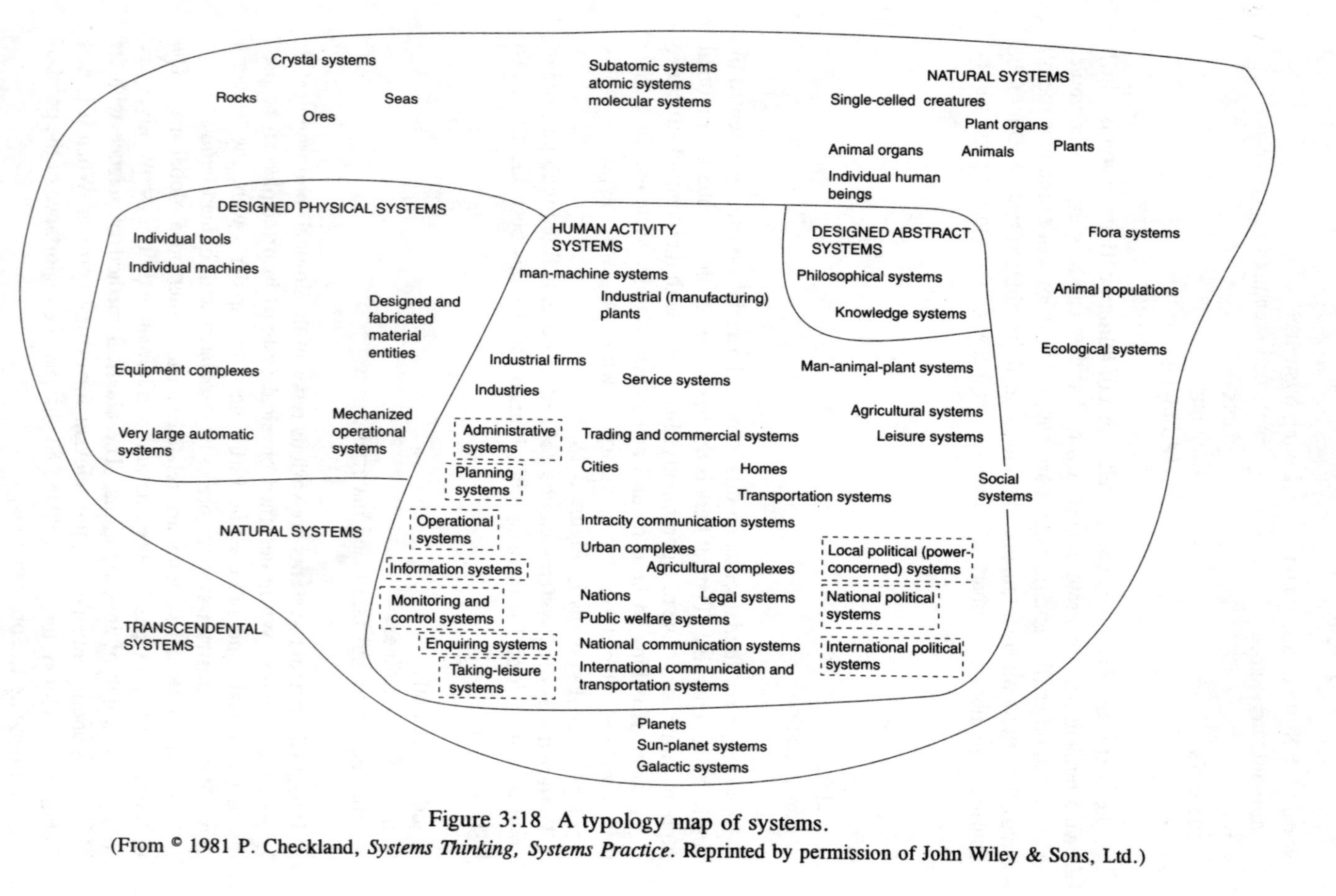

Figure 3:18 A typology map of systems.

(From © 1981 P. Checkland, *Systems Thinking, Systems Practice*. Reprinted by permission of John Wiley & Sons, Ltd.)

Jordan and the Systems Taxonomy

In an essay from 1968 published in *Themes in Speculative Psychology* an American psychologist, ***Nehemiah Jordan,*** presented his Systems Taxonomy. As a non-hierarchial structure it will only in part fulfil the conditions for being a general systems theory (see p. 32).

Systems taxonomy has three basic organizing principles which enable an observer to define a system as an 'interaction between what is out there and how we organize it in here': **rate of change, purpose** and **connectivity**. Each principle defines two antitheses of the other, thereby giving three pairs of properties.

Something which does not change within a defined time span (no rate of change) is *structural* or *static*; that which does change is *functional* or *dynamic.* Quite naturally, the actual time span determines which of the two qualities is relevant to use. In a very short time span, the dynamics are concealed, giving a static impression; in a very long time span, nothing can be static and the structure changes through entropy.

The organizing principle of purpose can generally have two directions: one towards the system itself and one towards the environment. Directed towards the system the aim is to maintain homeostasis. Directed towards the environment the aim is often to modify it to resemble a desired state or, if this is not possible, to bypass or override the disturbances.

According to Jordan, the concept of purpose is manifested by systems throughput. Every system whose input is internally processed and transformed to an output, is *purposive*. The output of the system is the desired goal; man-made systems are thus purposive. There are *non-purposive* systems as well: physical systems when in equilibrium (for example a volcano) provide an illustration.

Systems obeying the principle of connectivity can be assigned to either of two alternatives: the not densely connected or *mechanistic* and the densely connected or *organismic*. If an intervention into a system, with removal of parts and breaking of connections, produces no change of the remaining components, it is classified as mechanistic. In an organismic system the change of a single connection affects all others. A recapitulation is outlined in Table 3:4.

Table 3:4 The three organizing principles of Systems Taxonomy

Rate of change	Purpose	Connectivity
Structural	Purposive	Mechanistic
Functional	Non-purposive	Organismic

An analysis combining these principles and the pairs of properties gives eight alternatives. Jordan has arranged and exemplified these in the following order.

1 Structural Purposive Mechanistic	A road network
2 Structural Purposive Organismic	A suspension bridge
3 Structural Non-purposive Mechanistic	A mountain range
4 Structural Non-purposive Organismic	A bubble (a physical system in equilibrium)
5 Functional Purposive Mechanistic	A production line (a breakdown in one machine does not affect other machines)
6 Functional Purposive Organismic	Living organism
7 Functional Non-purposive Mechanistic	The changing flow of water as a result of a change in the river bed
8 Functional Non-purposive Organismic	The space/time continuum

The overall meaning of Jordan's Systems Taxonomy is to play down and simplify the often-misused concept of a system. Jordan states that 'the only things that need to be common to all systems are identifiable entities and identifiable connections between them. In all other ways systems can vary unlimitedly.' Concepts like feedback systems or self-organizing systems create more confusion than they solve and do not belong to his systems thinking. Finally, Jordan analyzes fifteen different definitions of the word system found in *Webster's New International Dictionary*. The result is that every one of the definitions is given its proper place in the taxonomy. Music, for instance as a system of sounds is defined in the dictionary as '(1) An interval regarded as a compound of two lesser ones – so used in Byzantine music. (2) A classified series of tones, as a mode or scale. (3) The collection of staffs which form a full score.' As a time-bound, functional system music fits alternative 7 above.

That music should be non-purposive highlights a major weakness of the taxonomy: the composer has a purpose, so do his listeners. Jordan ascribes the purpose (or its absence) to the system itself, leaving out its creator or observer. Should not a road network then be classified as non-purposive, just like a mountain range?

Salk and the Categories of Nature

Jonas E. Salk (1914-1995), immunologist and father of the Salk vaccine against polio, has proved himself to be a true natural philosopher in his book *Anatomy of Reality* (1983). Here Salk presents a number of conceptual maps showing how evolution as the primary cosmic force creates the order visible in the different categories of nature.

Evolution arises from the interaction of mutation and selection where mutation occurs by chance and selection by necessity for survival. This kind of true evolutionary change is characterized by its irreversibility. Salk uses the term evolution in a universal sense, as a force acting in the *prebiological*, *biological* and *metabiological* eras. These eras exemplify an evolution of the evolutionary process itself, with successively more sophisticated processes and strategies. In his words it is as if 'the principal preoccupation of evolution is its own perpetuation'.

In the categorization of nature, Salk begins with a definition of the three main eras of the universal evolution wherein the different types of matter emerged. See Figure 3:19.

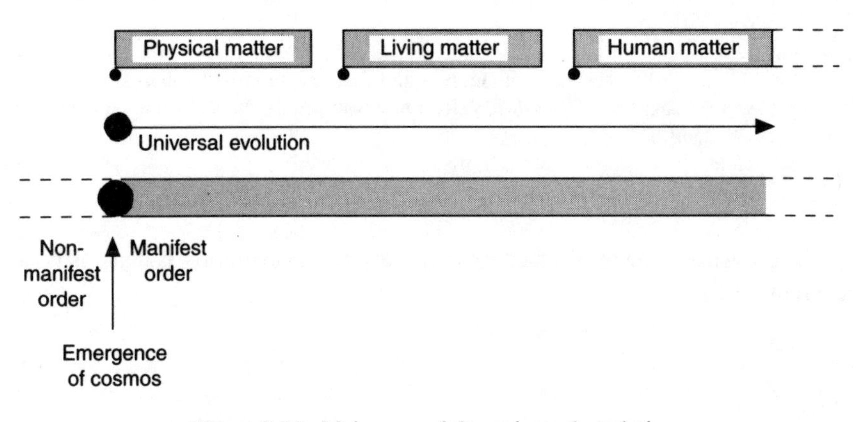

Figure 3:19 Main eras of the universal evolution.
(From © 1983 J.E. Salk, *Anatomy of Reality*. Reprinted with permission of Greenwood Publishing Group Inc., Westport, CT.).

The period of *physical matter* is prebiological; matter is successively condensed to a more complex structure. The period of *living matter* includes the foregoing period and is the same as the biological as life has now emerged. The *human matter* period, the metabiological, includes all prior periods and extends into the future. In this period matter can be said to become conscious of itself. Man has thus developed beyond all forms existing in nature but his full potential is still unknown. The human mind itself is considered to be a reflection of the surrounding cosmos, that is, contains the memory of the total previous evolution.

The era of the physical matter had the overwhelmingly longest duration; the emergence of each consecutive era took place at an increasing pace. The rise of order out of an initial chaos is also implicit in Figure 3:20. The evolution of a manifest order with the three main spheres is further demonstrated in Figure 3:20 below.

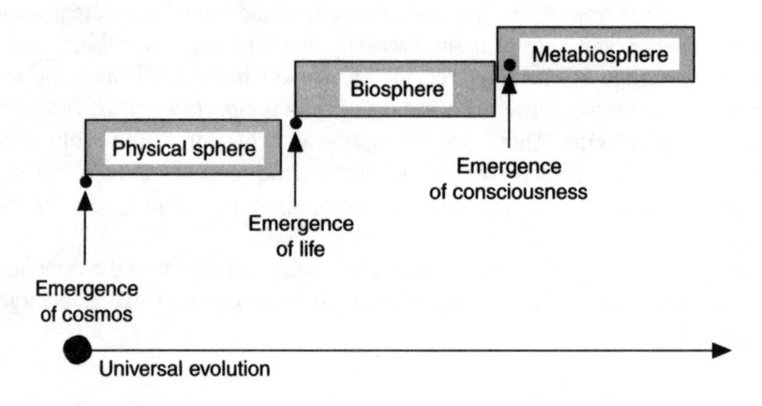

Figure 3:20 The three main eras and their evolutionary spheres.
(From © 1983 J.E. Salk, *Anatomy of Reality*. Reprinted with permission of Greenwood Publishing Group Inc., Westport, CT.)

The universal growth of complexity as part of this evolution is demonstrated in Figure 3:21.

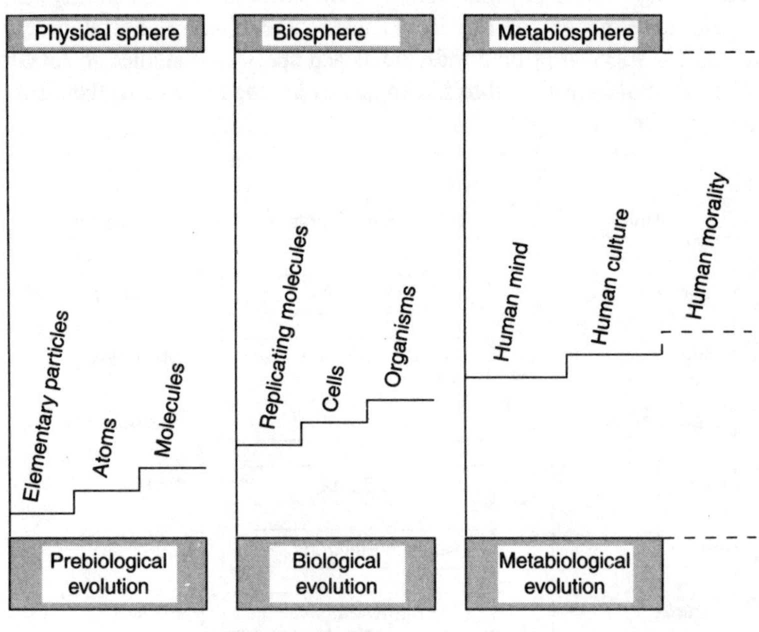

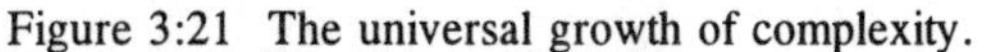

Figure 3:21 The universal growth of complexity.
(From © 1983 J.E. Salk, *Anatomy of Reality*. Reprinted with permission of Greenwood Publishing Group Inc., Westport, CT.)

The growing order and complexity has its first manifestation in the coherence of the elementary particles. A contemporary expression is the human mind and its culture, representing the highest degree thus far of complexity. The open-ended nature of the diagram suggests new metabiological stages, possibly a new kind of human morality or superconsciousness, as suggested by Teilhard de Chardin.

A more detailed order, called by Salk the *basic anatomy of reality*, is presented in Figure 3:22. Central in this diagram is the binary structure of the units at each level of complexity (for example, the relationship between energy and mass). According to Salk complexity has its origin in the tendency towards *complementary pairing,* within networks of pairs in functional relationships. In the right-hand part of the figure the various academic disciplines concerned with the different levels are shown.

This binary relationship joining inseparable factors is focused in the study of non-physical order in areas such as metaphysics, mathematics, philosophy, religion and art. Physical order *per se* is the focus of studies in physics, the origin and development of life in chemistry and biology. Social order, the conditions for survival of both individuals and species, is studied in sociology. Here the realm of human culture and creativity are the focus of metabiology and sociometabiology.

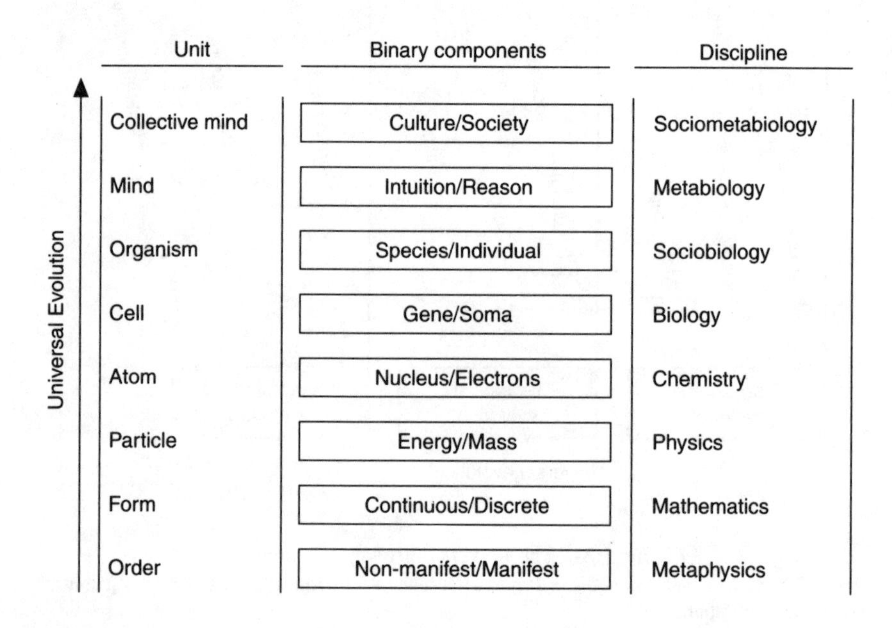

Figure 3:22 The basic anatomy of reality.
(From © 1983 J.E. Salk, *Anatomy of Reality*. Reprinted with permission of Greenwood Publishing Group Inc., Westport, CT.)

The concept of relationship is basic in the natural philosophy of Salk. It brings together the three dimensions of space and that of time building a fifth dimension of wholeness. If the fundamental units of the three main phases of evolution are defined as atom, cell and mind, a more detailed table regarding the critical *determinants of change* can be constructed. According to Salk these are *probability*, *necessity* and *choice*; they are represented by their corresponding attributes for each stage, namely *interactivity*, *procreativity* and *creativity*. See Figure 3:23.

Universal evolution

	Prebiological evolution	Biological evolution	Metabiological evolution
Emergence	Matter	Life	Consciousness
Unit	Atom	Cell	Mind
Components	Nucleus/Electrons	Gene/Soma	Intuition/Reason
Attributes	Interaction	Procreation	Creativity
Determinants	Probability	Selection necessity	Choice

Figure 3:23 The fundamental evolutionary relationships.
(From © 1983 J.E. Salk, *Anatomy of Reality*. Reprinted with permission of Greenwood Publishing Group Inc., Westport, CT.)

In his examination of the evolutionary process, from the beginning of the metabiological era, Salk found an ever-increasing pace in development. The mechanisms of mutation and selection in biological evolution have now been exceeded by human creation and choice, more rapid and efficient means of adaptation and of transmission to succeeding generations. Just as genes are tested or selected for their adaptive value, ideas have to be tested. An equivalent to the immune system within prevalent paradigms serves to preserve useful ideas. Human evolution will thus be determined in the future by the capacity for anticipation and selection and not by physical/biological mechanisms. Salk stresses the importance of *both/and* in critical decision making, where *either/or* has proved to be devastating. Salk's ideas regarding the supremacy of mental evolution may be compared with Teilhard de Chardin's nōogenetic evolution (see p. 73).

Salk sees evolution as both a problem-creating and a problem-solving process where new solutions grew out of the merging old solutions. But evolution has no preferences and only supports those species that can help themselves.

Another important aspect of the evolutionary process is its correcting capability. Also, imperfect systems with this capability can adapt and survive under all possible circumstances. Correctability implies mechanisms sensitive to feedback, as well as for feedforward or thinking ahead, necessary to effect the best choices. All this is held in the human psyche, designed by the development to correct mistakes in the past and present and to invent strategies for the future.

Powers and the Control Theory

As a psychologist and cybernetician, the American researcher ***William Powers*** had for many years pondered the question: 'Why does the same disturbance sometimes result in different responses?' The classical stimulus/response view and behaviourism had apparently failed to give an explanation. Powers himself gives the answer in his book *Behavior: The Control of Perception* (1973). Human behaviour is based on the concepts of control of reference perceptions and of feedback.

Powers illustrated his interpretation of these concepts with a small and ingenious metaphor: two rubber bands tied together with a knot. See Figure 3:24.

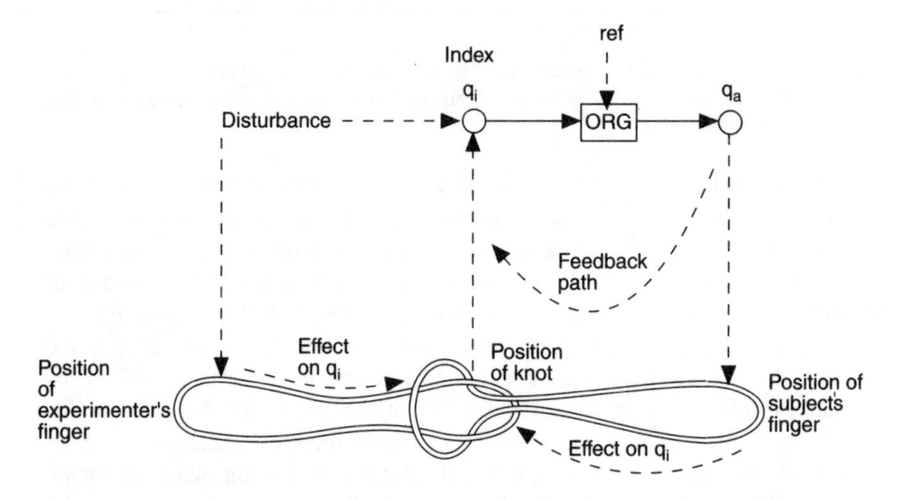

Figure 3:24 Illustration of feedback and control by use of two rubber bands. (Reprinted with permission from: Powers, William T., *Behavior: The Control of Perception*. New York: Aldine de Gruyter. Copyright © 1973 by William T. Powers.)

Two persons, A and B, standing on opposite sides of a table put a finger into the loop formed when two rubber bands are tied together. They stretch the rubber bands and adjust the knot above an index on the table. If A now starts a small movement of his finger to displace the knot, B reacts and is able to maintain its position. The position of the knot as seen by B and related to the index on the table, is the controlled quantity, q. The position of B's finger is the output quantity, Q. B's rubber band represents the environmental feedback path, whereby B's output affects his own input. The position of A's finger represents the disturbance and his rubber band represents environmental links through which a disturbance affects the same controlled quantity affected by B's output. Every aspect of the feedback control situation is thus both evident and explicit from the example.

Powers states that behaviour is governed by internal reference signals and that there exists a hierarchy of negative-feedback control mechanisms which are discernible in a person's behaviour. Within this hierarchy, the higher-level mechanisms set the reference conditions for lower levels and receive information about deviations in the comparison between controlled conditions and their reference values. Powers' hierarchy of feedback-control structures governs all kinds of human behaviour. This hierarchy of at least nine different levels is shown in Table 3:5.

Table 3:5 Powers' control hierarchy

Level	Core of control	Examples
First-order	Intensity	Muscle tension, spinal reflexes
Second-order	Sensation	Kinaesthetic perception
Third-order	Configuration	Posture, grasping, phonemes
Fourth-order	Transitions	Motion, time, change, warming
Fifth-order	Sequence	Walking, word sequences
Sixth-order	Relationships	Cause-and-effect, categorization
Seventh-order	Programmes	Looking for a pencil to write with
Eighth-order	Principles	Problem-solving heuristics
Ninth-order	System concepts	Perception of unities in abstraction
Higher orders	Spiritual phenomena?	

In the hierarchy only the first-order level interacts directly with the surrounding world. Neurological evidence of the proposed control levels exists up to the fifth order and Powers also indicates where it resides anatomically. Above the fifth, the different levels are less distinct and must be traced in a more indirect way.

In all hierarchies of control, the lowest-level system must have the fastest response. Adjusting reference conditions for lower-level systems in order to correct own-level errors must build upon the own slower performance and the performance of the lower system as well. Therefore, the higher the level in the hierarchy, the slower the adjustment and the longer the endurance of a disturbance. (While it is possible to see a clear correspondence between Beer's and Powers' hierarchies, on higher levels Powers' seems more speculative.)

Ultimately, the behaviour of an organism is organized around its control of perceptions. Perceptions have no significance outside of the human brain. A presumed external reality is not the same as the experienced. Even if we acknowledge a real, surrounding universe, our perceptions are not that universe. They are influenced by it but its nature and impact is determined by the processing brain.

Reference signals for natural control systems (see Figure 3:23) are established inside the organism and cannot be influenced from outside. In Powers' view a natural reference signal can also be called *purpose*. If a perception does not match its internal reference, the result is a perceptual error. The higher the hierarchial level of this perceptual error, the greater the psychological distress. Something has to be done to reduce the error, something which quite simply can be called behaviour. Therefore, in Powers' words: 'What an organism senses affects what it does, and *what it does affects what it senses*.'

A more practical application of Powers' model is found within the area of interpersonal *conflict*, *defence* and *control*. In Powers' view, a conflict is an encounter between two control systems which try to control the same quantity, but according to two different reference levels. A conflict is only likely to occur between systems belonging to the same orders; systems of other orders have other classes of perceptions. Hence no single controlled quantity is shared. Levels of orders other than those in conflict will therefore behave normally. The psychological concept of cognitive dissonance seems to be compatible with Powers' ideas with regard to this aspect.

A general model of the feedback/control system and the system's local environment sum up Powers' main train of thought in Figure 3:25. The nine levels are implicit.

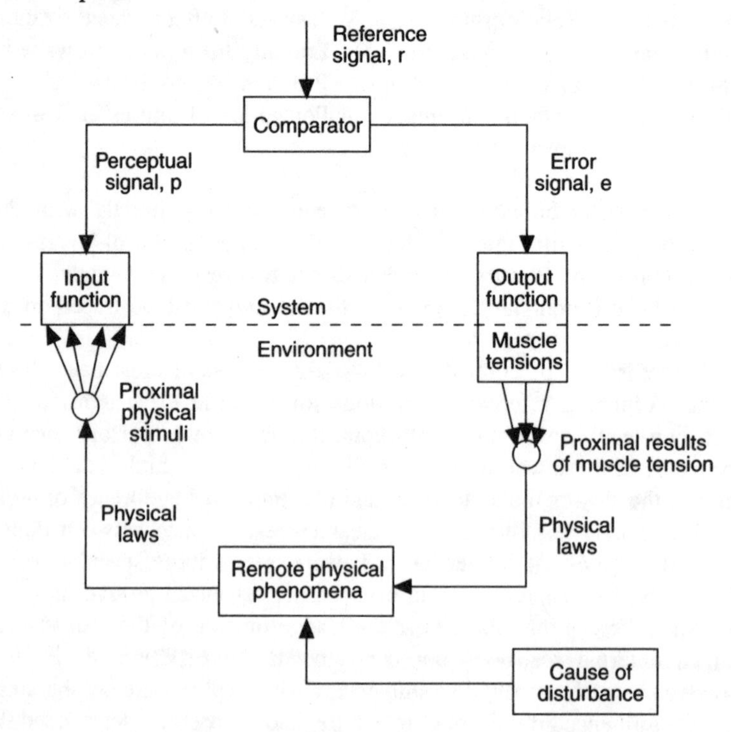

Figure 3:25 A general model of Powers' feedback/control system.
(Reprinted with permission from: Powers, William T., *Behavior: The Control of Perception*. New York: Aldine de Gruyter. Copyright © 1973 by William T. Powers.)

Namilov and the organismic view of science

Vasilii Namilov is a Russian researcher and philosopher of science. In 1981 he presented, in *Faces of Science*, a cybernetic approach to the phenomenon of science. In the book, the society of science is regarded as a metaphoric or abstract system, residing within the biosphere. As such it has the properties of a *macro-organism* obeying the same rules as other species within biological evolution. The book focuses on its self-organizing and self-regulating properties, with their equivalents in the biosphere. These properties in turn develop and change as the macro-organism evolves.

Namilov thus finds that the most typical property of both science and the biosphere is their organismic systems which develop over time. As the contained information is renewed, it is also complexified. Just as new species arise through the biological evolution, new ideas and areas of knowledge come into being in science.

The self-organizing system of science has its origin in the emergence of a communication system based on theses, textbooks, journals and other publications. The smallest component of the macro-organism, that is, the equivalent of a cell, is the scientific paper. Its development is determined by discoveries, by definition not possible to predict and thereby the equivalent to mutations. Through mutation, taken as a random generator together with some selection rules, an adaptation system is created – principally the same in the systems of both science and biology. An adaptation system always has a memory where new and useful information is stored. Namilov talks about a genetic memory in biological systems as well as a scientific memory residing in journals, books and libraries (see also page 167 regarding the different types of memories).

All information/communication systems have their own language. According to Namilov, while the genetic code is the language of biology, science has its own language for communication (not to be confused with conversational language). In this quite special language the information content of code signs used is constantly increasing. One consequence of this evolution is that scientific papers and articles become more compact, and increasingly incomprehensible for ordinary readers. Also, the exponential production of scientific papers has led to a publication crisis. Another consequence is the differentiation of science into disparate fields, each with its specific language. While such languages facilitate internal communication, external communication is aggravated. Here Namilov points out another analogy to biology: information structures of different species are incompatible.

Language itself is regarded as a natural organism in the eyes of Namilov. He refers to the words of the German linguist ***A. Schleicher*** (1888): 'The life of language does not essentially differ from the life of any other organism – plants or animals.' Namilov sees a struggle between the world of biology with its genetic language and the world of man with its semiotic language. The ecological crisis and the extinction of many species of plants and animals is the result of the struggle for survival of the texts written in these languages.

Changing external conditions in the biosphere, together with new experiments in science, exert pressure upon the internal information flow of the biosphere as well as of science. The consequence can be one of the following alternatives.

- A state of growing external information and of the generation of much new internal information exists. In the biosphere new species emerge and in science new theories arise explaining new, experimentally observed facts.

- While the flow of external information remains unchanged, internal information grows. In the biosphere original and often bizarre forms emerge and science degenerates into dogmatic and artificial constructs.

- The external information increases swiftly and the internal slowly. This will cause the dying out of species and a stalemate in science (as recently in elementary particle physics).

Together with the tendency to develop new forms and new ideas, there always exists a stability-maintaining mechanism in both of these macro-organisms. In biological systems, too pronounced variation soon results in maladjustment and the threat of extinction. In science, a new and revolutionary idea has to overcome the paradigmatic barriers before being accepted. In reality this implies a time constraint – no idea becomes common before the environment is prepared for it.

A basic strategy for all biological systems is that they transform their environment in a way that is favourable for themselves. An example is when plants take part in the decomposition of rocks in order to gain access to minerals. The same goes for science: in generating a favourable situation for itself, engineering developments create more instruments and apparatus, releasing in turn manpower for work in scientific areas. More scientists can join the pressure group and more efficient lobbying can secure more funds.

Another aspect with regard to the creation of maximum favourable conditions is its exclusion of other alternatives. Just as alternative life forms are hardly possible in the biosphere, Western-style science has monopolized the only accepted way to knowledge accumulation in the modern world.

Concerning the forming of species, both in the biosphere and within the scientific community, Namilov indicates obvious parallels. The important factors in both areas are conditions leading to isolation. In science, the elaboration of new ideas and pertinent information exchange take place within certain scientific schools, creating the necessary isolation. The phenomenon of symbiosis can also be recognized as mutual help with no internal exchange of information between different areas (for example cross-scientific research).

All information systems must get rid of excessive and outdated information. In the biosphere the organism, with its successively obsolete inherited and accumulated information, dies. Even whole species become extinct and are replaced by new ones when their information content no longer fits the surrounding world. In science, the old paper and the old book disappear giving space for new ones, something which can be traced in the citation bank. Disciplines may have different half-lives but normally this time span varies between five to ten years (who cites Kepler and Newton today?).

Finally, to complete his view of science as a self-organizing macro-organism, Namilov identifies the teleological traits of science. The search for the ultimate truth regarding man and the universe manifests itself in the development of ever-more complex hypotheses and theories.

Review questions and problems

1. Discuss which of the theories presented in this chapter arguably ought to have the greatest influence on the scientific community.

2. Which of the theories should be specially suited to be adopted in the management/business area?

3. Try to explain the practical implications of the Recursive System Theorem in a business organization.

4. In which of the theories is the concept of a nōosphere specially relevant?

5. The conventional evolutionary paradigm asserts that life adjusts to the environment. On this topic the Gaia hypothesis states a quite different view. Account for this.

6. What is the smallest entity known to physical science that deserves the name of a system according to Cook?

7. According to Namilov the life of a language does not differ from the life of any other natural organism. Report on his train of thought in this matter.

4 Information and Communication Theory

- **Information, exformation and entropy**
- **Shannon's classical theory**
- **How to measure information**
- **Entropy and redundancy**
- **Channels, noise and coding**
- **Applications of information theory**

'Information without communication is no information at all.'
(*Wurman* 1991)

Information theory is concerned with the problem of how to measure changes in information or knowledge content. It is based on the fact that we can *represent* our experience by the use of symbols like the alphabet, pictures, etc. But since the establishment of classical information and communication theory in the 1940s, nobody has succeeded in stating a general definition of the concept of information. As a result the commonplace definition where information is seen as the opposite to noise, or as non-meaning, is omitted.

Several conceptions of information exist, often used simultaneously and in the same context, which is a cause of confusion. The literal meaning of the word is as a rule 'that which determines form'. As commonly employed its significance is mostly derived from the context in which it is used. Information is neither matter nor energy, it is rather an abstract concept of the same kind as entropy, which must be considered a conceptual relative. 'Amount of information' is a metaphorical term and has in fact no numerical properties.

An existential view of information is relativistic and states that information *per se* is something imperceptible. Digital letters, numbers, sounds and images are a sequence of zeros and ones, not something possible to perceive as information. Pure information, like pure knowledge, signifies nothing at all; it is the context in which it is employed that gives it existence and value. Information becomes knowledge only when we decide to put it into use. Without this transformation, stored information is nothing more than physical or electronic signs.

Defined from a societal standpoint information may be seen as an entity which reduces maladjustment between system and environment. In order to survive as a thermodynamic entity all social systems are dependent upon an information flow. This explanation is derived from the parallel between entropy and information where the latter is regarded as negative entropy (negentropy). In more common terms information is a form of processed data or facts about objects, events or persons, which are meaningful for the receiver, inasmuch as an increase in knowledge reduces uncertainty.

Information scientists, however, with their more explictit need for clarity use some of the terms defined below to measure information content. ***Selective information*** has to do with with the number of minimum independent choices between two equally likely possibilities. This gives the promise of narrowing the range of prospects about which we are ignorant. Its measure is a relation between a signal and an ensemble. ***Descriptive information*** is seen as small entities which, when added together, build up more knowledge about something. A microscope with higher resolution accordingly gives more *structural* information about an object than one with lower resolution. On the other hand, when an observation gives more precision by better instrumentation and finer readings (more decimal places), it has gained in *metrical* information content. The measures of selective, structural and metrical information-content can be seen as complementary. A simple analogy of their mutual relationship would be volume, area and height as measures of size.

Information is always dependent on some physical base, or energy flow, where the energy component is subordinate to the structure of variation, manifested by the flow. The structure of variations in the media used must always remain unaffected by the carrier, however it is chosen. If these variations in some way match the structure of the receiving entity, a dynamic relation is possible. Information is therefore a kind of *relationship* between sets of structured variety and not a substance or concrete entity.

Information has also been defined according to what has been called the *infological equation* (Langefors 1973). It is represented by the following formula:

$$\mathbf{I} = \mathbf{i}\ (\mathbf{D}, \mathbf{S}, \mathbf{t})$$

In this formula, **I** stands for the information achieved by an interpretation process, **i**, acting on data, **D**, with regard to previous knowledge, **S**, during the time, **t**, which is available.

Whatever the definition, information is an invisible agent – such as electricity in a modern town, tying together all components (personnel, machines, money, material, etc.). Decision making and control, regulation and measurement are affected through information. This has therefore to be filtered, condensed, stored, transmitted, received, aggregated, integrated, manipulated, and presented. These processes are based on the fact that information may be infinite but only possible to organize in a finite way from a human point of view. All information can be structured according to the following:

- Category
- Time
- Location
- Alphabet
- Continuum

Of course, each choice has many variations but the main alternatives are still basically five in number. If we take as an example a book going to be part of a library, this demonstrates that when a structure is used, the book is easily recognized.

Category in a library means the main topic according to the content of the book. Such topics are fiction, history, philosophy, etc. Time in a library sense is the printing year of the book, while location is its physical position on one of the shelves. Alphabet is the arrangement of the book stock in alphabetic order, both with regard to the author's name and the title. Continuum is the current newsletter presenting all recently acquired books in the library stock. Traditionally, libraries have used all structures simultaneously, well aware that each way of organizing information will permit a different understanding. The possibility of multiple perspectives is a good approach when the aim is to extract maximum value and significance from information.

Turning now to communication, a more concrete term, this applies to all sorts of transference in one or more directions of matter (as objects), energy or information. It may include the conveyance of written, acoustic, visual or other kinds of messages. The problem of communication can be formulated as: making a representation in one place of a representation already present in another place. A behavioural definition of communication offered by ***Warren Weaver*** (1949) reads: 'All of the procedures by which one mind may affect another.' Without losing its validity, this definition can similarly be applied to communication between machines.

Comunication is usually transference of **representative substitutions** for that which we want to communicate. These substitutions are carried by **signs** consisting of signals and/or symbols. A sign is everything which can be taken as adequately substituting for something else. Most often the sign seems to be more easily comprehended than the thing it signifies, while the thing itself is more difficult to comprehend.

The **signal** is most often a physical phenomenon in the environment which has a direct relation to its object and is itself a part of what it represents. It has no need for active interpretation and supplies its message directly. Red rockets launched at sea is an emergency signal understood by all seamen. Signals may be transmitted in time, for example as speech, or in space as print or pictures. A **symbol** is an internal or mental representation, referring to its object by a convention and produced by the conscious interpretation of a sign. In contrast to signals, symbols may be used every time if the receiver has the corresponding representation. Symbols relate to feelings and thus give access not only to information but also to various kinds of emotions.

Communication originated from the transference of objects and later came to be involved with the handling of signs representing the object. Ultimately, communication handled symbols representing the signs. Communication theory is mainly concerned with the processes by which messages can be coded, transmitted, and decoded. Communication and information theory are closely linked.

Analyzed in terms of information theory, communication is categorized into two different methods, the analogue and the digital. The **analogue** method of communication has no either/or, instead it represents both/and. It is an ongoing continuum and symbolizes an analogue or icon of a real process and as such it is a process of relationships. **Digital** communication represents the choice between either/or and is a reproduction of structure and pattern.

Natural language is characterized by an openness, permitting all kinds of new messages and represents the most advanced form of digital communication. Here the words represent fixed points for concepts, and to manipulate the words is to manipulate the concepts themselves. All living systems, however, employ both analogue and digital methods in form and function at some level of the communication system. Verbalization and symbolization include the digitization of the analogue. In human communication the translation from analogue to digital often results in a gain of information but a loss of meaning. Alteration between the two methods often takes place when communication crosses certain types of boundaries.

An illustration of the differences between the analogue and digital mode is presented in the following list of opposites:

Analogue mode	Digital mode
– Continuous scale	– Discrete units
– Concrete	– Abstract
– Territory	– Map
– More/less	– Either/or
– Continuous	– Discontinuous
– Presence and absence	– Presence or absence
– Relativistic	– Absolutist

Communication theory covers a broad scope of phenomena emerging from communication in various kind of systems. Examples of phenomena to investigate are qualities like detectability, localizability and informability in living system communication (Skyttner 1993).

Information, exformation and entropy

Apart from the above-presented perspectives on information, two views are predominant among researchers. The **mathematical/statistical** view is primarily used in connection with telecommunications and databases to quantify and measure channel and storage capacity for information exchange and processing. This concept has no epistemological aims to explain the nature of information. It was used by Shannon and Weaver in their pioneering work (see the following pages). Here information is defined as a measure of freedom of choice when selecting a message. Their concept of information must not be confused with meaning. Meaning can be defined as the significance of information for the system which processes it and has to be measured by units other than those normally used in information theory. Meaning does not relate to the symbols used, rather to what the symbols represent.

The **information physics** view states that information is an implicit component of virtually every equation governing the laws of physics. Information is a property of the universe and it does not have to be perceived, to have meaning or to be understood, in order to exist. The information physicist ***Tom Stonier*** (1990) has published some theorems concerning the interrelationship between matter, energy and information. Some may be regarded as quite revolutionary by classical information theorists. Freely interpreted they are as follows:

- All organized structures contain information, and – as a corollary – no organized structure can exist without containing some form of information.

- The addition of information to a system manifests itself by causing a system to become more organized, or reorganized.

- An organized system has the capacity to release or convey information.

- Heat is the product of energy interacting with matter. Structure is the product of information interacting with matter.

- The information content of a system is directly proportional to the space it occupies.

- Time, like entropy, is inversely related to information. The greater the interval between two events, the less the information content of the system.

- Energy may be converted into information or be used to convert information from one form to another (information transducing).

- The more highly organized a system is, the more its information is separated from the energy that bears it. (Information should not be confused with the matter-energy markers which bear it.)

Organization may thus be expressed as the manifestation of information interacting with matter and energy. When added to matter it exhibits itself as structure or organization; consequently organization may be regarded as stored information. Information may not only organize matter and energy, it may also organize information itself. From that it follows that the more organized a structure or process is, the less information is needed to describe it completely. On the other hand, a loss of organization is always associated with an increase in entropy. Furthermore, information is everywhere, but knowledge only exists within a goal-seeking adaptive system consisting of goal-seeking subsystems.

A modification of the information content of a system always results in a corresponding change of its pertinent entropy. Such a change may be brought about through the alteration of its organization or its heat content. The function of additional heat may be to stabilize the organization and minimize externally induced entropy. If all existing heat were withdrawn from a system, its temperature would be zero degrees Kelvin and also its entropy would be zero. A phenomenon such as this is defined by what has been called the **third law of thermodynamics**.

The expansion of a system in physical space as a result of the application of energy will not produce change in information, unless accompanied by change in organization. The information content of a system normally tends to vary directly with the space occupied and inversely with the time occupied. Concerning the impact of time, physical information is time-dependent and does not withstand the forces of entropy. A system which is more resistant to erosion over time has to contain more information. The configuration of a physical system blurs with time and consequently an observation of it becomes increasingly obsolete.

Several inventions within the area of human communication enact information patterns on various forms of energy. Radio transmitters and printing presses, for example, multiply the information content several million times.

Accepting this view of information-physics gives some inevitable general consequences for the relation between **information, probability** and **entropy**. Information, order and improbability create the opposite to lack of information, disorder and probability, which represents entropy. Information is therefore to be considered the *inverse* of entropy and is sometimes referred to as negative entropy (or **negentropy)** by some researchers.

These facts explain why the famous Maxwell's demon (see p. 12) does not disobey the second law of thermodynamics when it sorts molecules through a trap-door in a closed system. By his decisions about energy levels the demon generates negentropy, the local increase of which is necessarily matched by more entropy elsewhere.

A paradox revealed by information physics is that as the universe evolves, its information content increases and it may end up in a state where all matter and energy have been converted into pure information. The main laws of thermodynamics state that the total entropy in the universe has to increase. Evidently there exist simultaneously two universal contradictory forces: one entropic, destroying and levelling out, and one organizing and building up.

Another paradox is that meaning and information have very little to do with each other, just as no directly visible correlation exists between order and information. The more disorganized and unpredictable a system is, the more information it is possible to obtain by watching it. One can never know if a hidden order exists; it may well exist even if it is not possible to reveal for the moment. Organization, information and predictability are thus quite paradoxically interrelated.

A closer look at the concept of information will reveal some strange qualities. One is that it is impossible to wear out information. You may duplicate information in as many copies as you want without deterioration of the source and moreover mostly without cost. To get rid of information will on the other hand always involve some cost.

Information may be measured on the basis of the amount of surprise, unpredictability or 'news value' that it conveys to the receiver. Paradoxically, disorder possesses a greater surprise potential than order. A completely unstructured sequence of letters like EVSYEDTOQPF is very difficult to describe; there is no simpler description than the sequence itself. It must therefore be assigned a maximum of information. A structure imposed on the letter source will reduce the average amount of information per letter from that source. Whereas an ordered row of letters, say ABCDEFGHIJKLMN, holds less information, the combination EINSTEIN can provide a great deal of information. The latter row has been the subject of more information processing in the ordering of the letters in relation to a meaningful context. Its information history is composed of the knowledge of both the humanity and the whole Western scientific culture. Information is extremely context dependent and very often the real content of a message is read between the words.

The reception of information will normally result in a decrease of uncertainty for the receiver. This process is shown in Figure 4:1. The slope of the channel conveying the information shows a linear relation by the 45° angle.

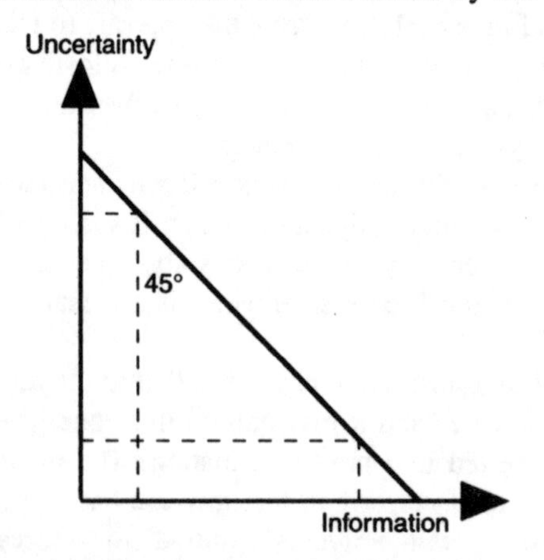

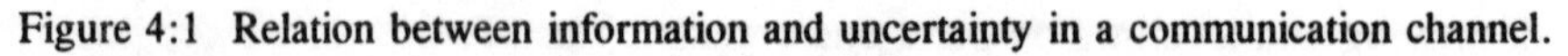
Figure 4:1 Relation between information and uncertainty in a communication channel.

The amount of uncertainty reduction is dependent upon the amount of information already held by the receiver. According to this pre-existing information the slope in the figure may have a quite different angle.

The communication system with transmitter, channel, receiver, etc. and its given set of message categories constitutes a closed system. Information expressed in categories other than those normally used therefore has a tendency to be interpreted as an error arising from distortions or mistakes. Information, however, is normally ordered simultaneously on many levels. The regular effect of communicated information is surprise followed by uncertainty reduction. Information which initially increases uncertainty evokes a higher order of surprise related to a discontinuity in the information accumulation.

Imagine the optical information system of traffic lights. Each driver knows what to do when the light shifts to green via amber, but what if the light suddenly turns blue instead of green? With a shift from one level of categories or channels to another the reduction of uncertainty ceases. When the passenger tells you that blue light is a request to pull over and stop the car in order to let the fire brigade pass (exclusive for this town), you say 'Aha!'. A fresh set of categories accommodating both the old and new is established and the uncertainty reduction can be resumed. Figure 4:2 depicts graphically the discontinuity in uncertainty reduction and the transition to a more diverse channel, all related to the example given above.

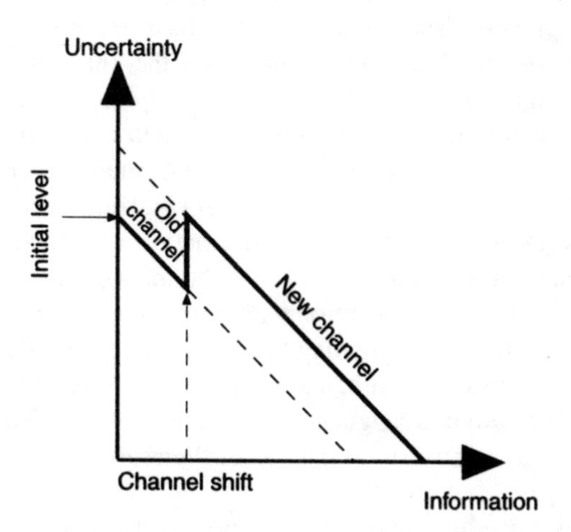

Figure 4:2 Discontinuity in uncertainty reduction and transition to a more diverse channel.

A discontinuity is also possible in a reverse way. Imagine the car at the red light when it suddenly turns green without showing amber first. This transition to a less diverse channel poses no problem as the information received still fits into the basic content of the existing category and need not be reinterpreted into new categories. Both red and green retain their meaning (see Figure 4:3).

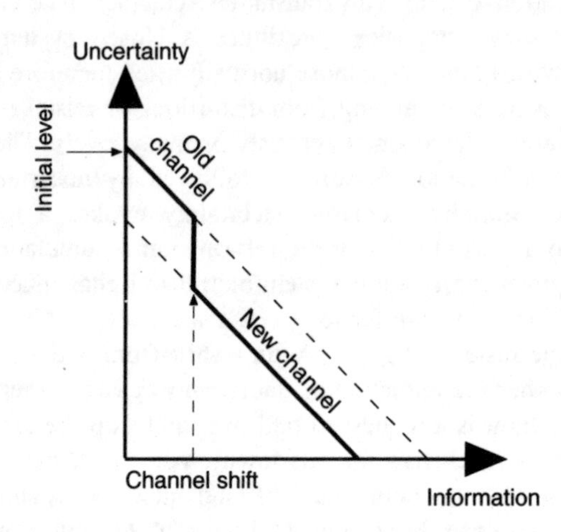

Figure 4:3 Discontinuity in uncertainty reduction and transition to a less diverse channel.

The shift between channels of varying complexity is an important part of the communication process. The involved parties have to ensure that the channels used are neither too simple nor too complex for the kinds of message used and they must adjust them as required.

A message may hold information which, although not present in the message *per se,* is comprehended by the receiver through reference to previous known facts. This is the basis for the concept of **exformation,** derived from external information (Nörretranders 1993). It is information which exists in the head of the sender, is omitted in the composition of the message and is presumed to be deduced by the receiver. The intention with a message is primarily to induce the receiver to form an idea corresponding to that of the sender. This use of exformation is possible because human beings share experiences which are possible to interpret through a common language giving the same associations. The words refer to something not intrinsic to the words themselves but conceivable in the mind of their user.

Information is something measurable, expressed by letters or bits used in the concrete message, while exformation is all that was omitted or extracted. More information is not necessarily more exformation. It is therefore not possible to measure the amount of exformation; this is dependent upon each context. In certain circumstances the omission or non-existence of a special signal may stand for a complete message. A phrase like 'silence speaks volumes' tells us that a general silence may convey a very comprehensive message. Thus the information content of a message clearly depends on someone's prior expectations about the message. With this perspective we are not able to speak of how much information a person has, only how much a message has.

Empty spaces within the organized structure of a message may also be highly significant pieces of information. The most frequent symbol in written English is the space between words which conveys information until the words are removed and the page becomes blank. It is the organization and structure of the surrounding system which defines the information content of existing empty spaces. Spaces as discontinuity define boundaries of structural entities, but the absence of structure within a structure can sometimes constitute information as significant as the structure itself. The value of information may here be defined as the amount of work which is performed by the sender and which the receiver need not repeat.

Shannon's classical theory

In his paper from 1948, *The Mathematical Theory of Communication*, ***Claude Shannon*** presents the foundation of classical communication theory and differentiates between three conceptual levels: the syntactic, semantic and pragmatic. As we will see, these levels can be adapted to all main concepts within information theory, such as noise, redundancy, etc.

The **syntactic level** deals solely with the internal relation between signs used. This includes rules for building up sentences in a formal manner. The **semantic level** is the level of application and of general understanding of the significance of signs used to relate to things, actions and the outside world. The **pragmatic level** is the level of the user world, of the personal and psychological impact of communication. Here questions of meaning, results and value for both sender and receiver are actualized. If human links in the chain of communication are missing, of course no questions of meaning will arise.

These three levels may be exemplified by the three stages in handling a spoken language, implying two intermediate translations between the outside world and the subjective reception of information. The first level consists of the acoustic pattern taken physically as vibrations in the air. The second level consists of the various phenomena in the inner ear and pertinent parts of the nervous system. The third represents the conversion of the information pattern by the brain into an experience of individual meaning.

The three conceptual levels of communication by Shannon are transformed into three related problems. The first is the **technical problem**: how accurately can the symbols used in communication be transmitted? The second is the **representation problem**: how accurately do the transmitted signs represent the intended message? The third is the **efficiency problem**: how efficiently does the received message influence the behaviour of the receiver.

In his work, Shannon explicitly states that the presented theory relates only to the syntactic level and the technical problem. The theory thus concerns the probability of the reception of certain signs under various conditions in the transmission system. Information is an entity regarded as neither true nor false, significant nor insignificant, reliable nor unreliable, accepted nor rejected. The coding of experience into a set of communication symbols and its recall after transmission by decoding, the very content, is irrelevant and outside the scope of the theory.

Although the representation and efficiency problems are irrelevant for the technical problem, the technical problem is highly relevant for the representation and efficiency problem. All calculations of representation and efficiency are dependent upon the precision in the technical. Whatever its form, the message has first to be received properly before the content can be perceived.

In Figure 4:4 the main concepts of Shannon's theory are presented. The theory is completely general and the communication process is seen as a transaction between terminals with the sole aim of generation and reproduction of symbols. It may be applied to a person communicating to himself (writing a memo) or to an unintentional interceptor of a message. If humans form parts of the communication system, the mathematical theory is relevant only in the technical part of the system.

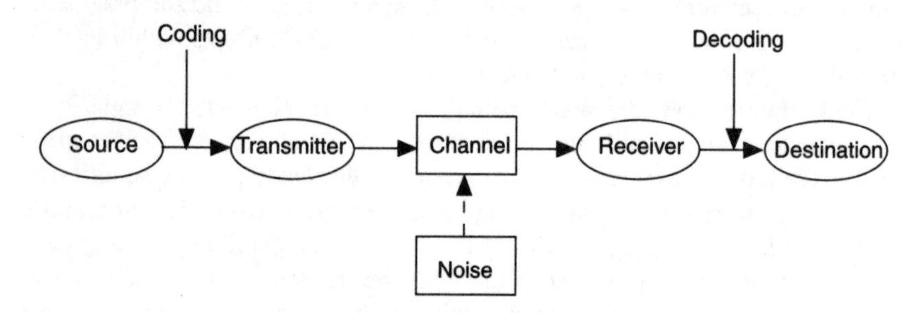

Figure 4:4 Concepts of Shannon's communication theory.

Every message intended to be communicated to someone has a **source**. The **message** which consists of a sequence of **symbols** from a certain repertoire is forwarded to the **transmitter** and then sent through a **channel** to the **receiver**. The receiver is connected to the **destination**. In this process the information is first **coded** and then **decoded**. It is thereby assumed that both the source and the destination have previously agreed on a code with the same or similar referents, used in such a way that the messages have meaning and relate to behaviour.

Both transmission and coding take advantage of the spatial, temporal or other classificatory ordering of the elements or *markers* carrying the information. **Noise**, which is always present in the channel, interferes with the transmission, degrading its quality to a greater or lesser extent. The total influence of noise can be measured as the resulting difference between the input and the output message.

In contrast to Shannon's classical theory, modern information theory makes a distinction between *channels of communication* and *channels of observation*. This is based on the fact that communication requires at least two persons, while observations and measurements require only one (nature does not communicate to us with language or signs).

In the communication system presented in Figure 4:4 some critical conditions must be fulfilled for optimum performance.

- According to its intentions, the information source must provide adequate and distinct information.
- The message must be correct and completely coded into a transmissible signal.
- Taking into account different kinds of noise and the needs and aims of the destination, the signal has to be transmitted in a sufficiently rapid and correct form.
- Received signals must, in spite of disturbances, be translated into a message in a way that corresponds to the coding.
- The destination must be able to convert the message into the desired response.

An examination of the communication itself during the transmission of a written message will show a general hierarchical structure of the process. Imagine the short electronic message, 'remember today's meeting' mailed from one computer to another. To analyze this message, four levels will be used according to Figure 4:5.

Pressing the letter **R** on the keyboard activates a certain binary pattern generated by the computer, here defined as the zero level. This pattern is assigned to a particular letter or keyboard function, now interpreted on the screen as the letter **R** on the next level one. When additional letters are formed they build the first word on level two. Level three consists of complete sentences, eventually building a paragraph at level four (see Figure 4:5).

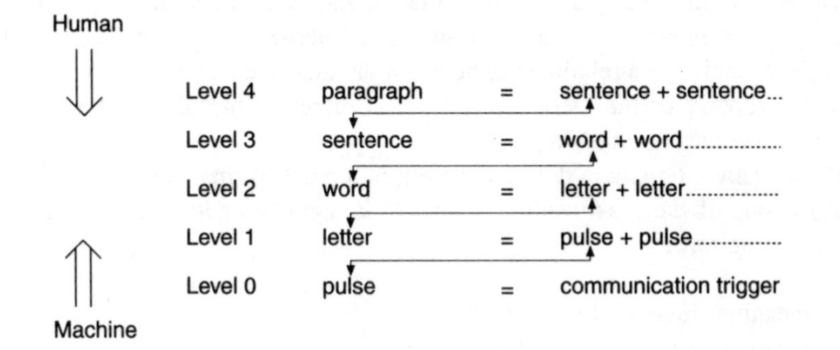

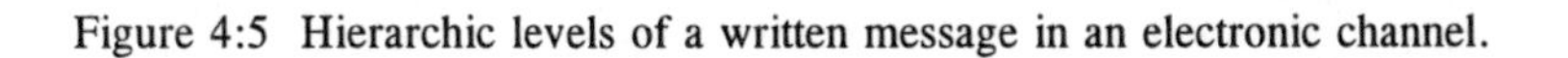
Figure 4:5 Hierarchic levels of a written message in an electronic channel.

Being occupied only with the first two levels of the hierarchy, Shannon's theory concerns the binary pattern representing the letters (coding). Effects of intervening noise occurring in the channel and the application of redundancy to neutralize the subsequent existing errors are also dealt with. The hardware is totally indifferent to the choice of letters used. The combining of these letters to create words is directed not by the hardware but by the rules of the language and the linguistic customs of the human user who wishes to be understood. Words are then arranged into sentences according to the syntax of the language and these in turn may be further arranged in paragraphs to emphasizing and distinguishing different topics. Several prelinguistic processes have however been completed before the specific message is chosen and typed out. The origin of human communication begins on levels far higher than those existing in Figure 4:5. The move from the highest descriptive to lower levels clearly shows how certain systematic changes occur.

Given his background as an electrical engineer, the restriction of Shannon's work to the technical problem is natural. He must, however, have been aware of the extreme difficulties present in the development of a theory for the other two levels. A complete theory of communication must deal with the structure of the message and its encoding, communication, decoding, and understanding, of the inherent meaning. Finally, the theory must calculate how efficiently the message will contribute towards desired behaviour.

As we now understand it, classical information theory tells us as little as does the hardware used about the incorrect choice of words or use of a vague structure in a sentence. A vague structure may even be intended by the source and it should be possible to transfer every kind of message, even sheer nonsense (genuine or as used, for example, in cryptography). Shannon's concept of information therefore excludes meaning *per se*; what is significant is only that the actual message is selected from a set of possible messages. How the source chooses its messages is irrelevant to the theory.

Although the concept of meaning lies outside formal information theory, the pragmatic aspects influencing information must not be forgotten. The concept of information is thus inseparable from that of meaning. Meaning is always the meaning for *someone*, defined in terms of the person or system receiving the message. Meaning is a relationship between the message and the receiver and no inherent property of the message itself. It is therefore meaning that can differ from one receiver to another.

Of the many attempts to define meaning in a more formal way, ***D. MacKay***'s (1969) is one of the most useful. MacKay distinguishes between the following aspects of the term:

- the meaning intended by the sender
- the meaning understood by the receiver
- the common meaning

He then defines the meaning of a message as its selective function on a specified set of responses, something valid for all the three aspects.

Every message is accompanied by certain implicit instructions for the receiver as to how to interpret it and relate to the sender. This kind of *metamessage* is always superior to the pertinent content of the message and is generally conveyed analogically, for instance by intonation, facial expression or bearing.

C. Cherry (1966) points out that the pragmatic qualities of a message are dependent on:

- Earlier experiences by the sender and receiver
- Present circumstances
- Individual qualities

Whereas Shannon's theory mainly belongs to the area of communication, he presents a pragmatic view of the highly abstract concept of information. The transitory nature of information has traditionally made it an integrated part of the media used. Separating information from its material carrier is nevertheless a prerequisite for the understanding of its nature. Information is not dependent upon any specific technology for production, distribution and use. Only when separated from its technology will information become an adequate measurable entity.

How to measure information

If information is to be treated scientifically rather than philosophically it has to be expressed numerically and quantitatively and not be interpreted as synonymous with meaning. Much confusion is caused by attempts to identify meaning with the change generated in the receiver. What is actually sent is not a measure of the amount of information transmitted. This depends merely on what could have been sent related to the prejudiced view of the expected message.

The information content of a message is nevertheless related to the complexity of its structure. An extensive initial lack of knowledge by the receiver gives a high complexity of the message. Structural complexity of the message thus may be used to define the quantity of information contained in the message.

As defined by information theory, the concept of information is merely a *measure of the freedom of choice* when selecting a message from the available set of possible messages formed by sequences of symbols from a specific *repertoire*. In reasonably advanced communication systems this set of possible messages may be formed using aggregates of words, e.g. in the English vocabulary. Furthermore, the system must be designed to convey every possible selection, not only that one selected at the moment of transmission.

The information content can be determined functionally using the difference between the initial uncertainty before receiving a message and the final uncertainty after receiving it. In this way we get a working definition of information as being the amount of uncertainty which has been removed when we get a message. The information content when receiving a message in one phase is:

Initial uncertainty – final uncertainty = Total information

If received in two stages, the information content is defined as the difference between the initial uncertainty and the intermediate uncertainty.

Initial uncertainty – intermediate uncertainty = Initial information

When further information is added, the final information is defined as the difference between the intermediate uncertainty, plus the difference between the intermediate uncertainty and final uncertainty.

Intermediate uncertainty – final uncertainty = Final information

Quantitative relationships in such stages may easily be added together. In the above simple equations the terms intermediate uncertainty cancel each other. Total information again equals the difference between the initial and the final uncertainty. In this way the information quantity assigns a value to the content which describes the complexity of the message and which may easily be added.

Information may be measured in terms of decisions and its presence can be demonstrated in reply to a question. The question is posed because of lack of data when choosing between certain possibilities; the greater the number of alternatives, the greater the uncertainty. The game of Twenty Questions illustrates how an object is supposed to be identified through answers to questions concerning the object ('yes' or 'no' questions). The strategy of uncertainty-reduction in the game is easy to recognize in Figure 4:6.

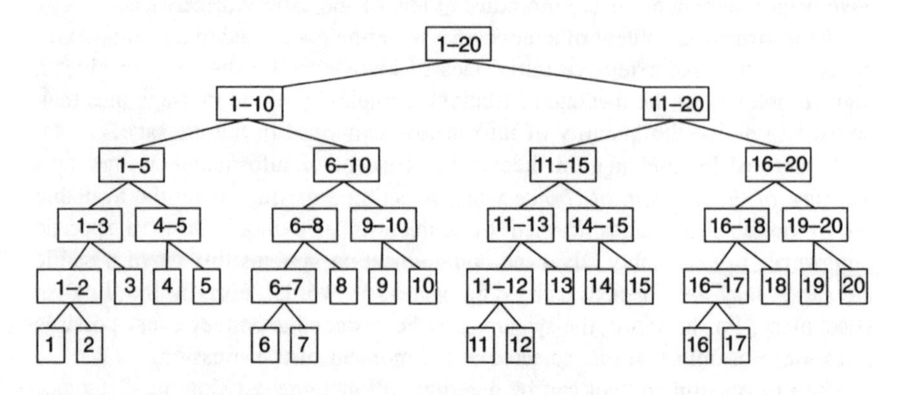

Figure 4:6 Strategy of uncertainty-reduction in the game of Twenty Questions.

Let us start with a situation where we can pose a single question: is the new-born baby a boy? Here it is equally possible that the reply will be yes as no and when the reply is given no uncertainty remains. The structural complexity and the information content are thereafter the smallest possible. The quantity of information contained in the answer may be defined as one unit of information. In information theory this is more precisely called one 'bit' of information. The quantity is derived from the repertoire of the digits 0 and 1 in binary notation, both assigned equal probability and carrying a content of one bit. Each question must divide the field of possibilities equally if one bit of information is to be gained for each reply.

With two questions, one out of four possibilities may be decided; with three questions one out of eight and so on. It is obvious from the examples that the base 2 logarithm of the possible number of answers can be used as a measure of information. Eight possibilities gives log 2 for 8 which is 3. Thus we have here three bits of information. With only one possibility there is evidently no uncertainty at all; the amount of information is zero, as the logarithm of 1 is zero.

The base 2 is a very natural choice because it comprises the minimum number of alternative messages in the repertoire of even the most primitive communication system. An example of such a system is the old-time acoustic fire alarm given by a special tolling of church bells: ordinary tolling = no fire, special tolling = fire. It is however unrealistic to expect equal probabilities regarding the messages contained in the system's repertoire: the no-fire situation is more probable than the fire. The lack of parity between different messages in a possible repertoire therefore corresponds to the probability of their selection. The proposed measure of information ($\log_2 2$) must therefore be revised to include preferences for a certain kind of choice.

To communicate a message is to transmit a pattern distributed in time, seen mathematically as a time series. A measurement of information is principally related to the regularity or irregularity of that pattern, but the irregular is always more common than the regular. A random sequence of symbols shows no pattern and conveys no information. A fundamental principle of information theory is therefore that information is characterized by symbols with associated probabilities of occurrence. Written language, for example, is a source where the symbols used appear with unequal probabilities and are statistically linked. Its elements are therefore mainly discrete, separate and mutually exclusive.

The amount of information carried inside a message is determined by its probability in the set of all possible messages. The more probable the type of pattern, the less order it contains, because order has less probability and essentially lacks randomness. It is therefore obvious that the less probable a message is, the more meaning it carries, something intuitively felt. We apprehend the surrounding world not on an equal scale of probability, but on a scale which is heavily biased towards the new and interesting.

Probabilities are by default always less than or equal to 1, because 1 is the probability of absolute certainty and no probability can be greater than absolute certainty. From that it follows that the amount of information is determined to be greater than zero when the probability of the matching event is less than one. If the selecting probability in a set of messages is 1.0, the message is always chosen; no freedom of choice exists and no information is communicated. To conclude: when probability approaches zero, information approaches infinity; when probability approaches 1, information approaches zero.

While information combines additively, probabilities taken independently combine multiplicatively. Thus the relation between the amount of information existing in a message and the probability of that message will be similar to the relation between a set of numbers that multiplies and a set that adds. Mathematically, the first set is defined as the logarithm of the second set, taken to an appropriate base. The handling of logarithms however demands a suitable scale, determined by a factor, positive or negative, by which it can be multiplied. A mathematical property of information conveyed by an event occurring with a certain probability, is that its probability logarithm has a negative value. The ordinary logarithm of a quantity less than one is always negative, while information is naturally taken to be positive. By adding a constant quantity it can be made artificially positive – a result also given if starting from a value other than zero.

The measure of freedom of choice, i.e. the information content, in a repertoire of two messages with different probability may be exemplified by the already mentioned tolling of the bell. If every tenth tolling of the bell on average is a fire alarm, it must be assigned a probability value of 0.1. All other tolling sequences thus comprise the probability of 0.9. Then the information content of a tolling may be calculated as:

$$-(0.9 \log_2 0.9 + 0.1 \log_2 0.1) = 0.476 \text{ bits}$$

When a particular message becomes more probable than any other, the freedom of choice is restrained and the conforming information measure naturally has to decrease. The information content of the message repertoire does not depend on how we divide the repertoire. The content of each individual message can be computed individually and then added to form the total message.

In applications of information theory it is practical to consider the letters of the alphabet as the repertoire of available messages (each letter being a kind of elementary message). The 26 letters of the English alphabet and the space needed to distinguish between words give a total of 27 symbols. Equal probability of all the symbols in this communication system would assign them the same individual information content of $\log_2 27 = 4.76$, i.e. 5 bits. (Note that every bit represents a choice between two alternatives and that it is inconceivable with fractions of choices here.) In reality, however, they occur with very different probabilities so the average information content of an English letter is about 4 bits.

When utilized in a real message the information content of a letter is still lower, as the English language inherently restricts the freedom of choice. Constraints induced by grouping and patterning of letters, words and compulsory redundancy cause the real information content to be a little less than 2 bits per letter. The probabilities of occurrence of certain pairs, triplets, etc. of letters are astonishingly constant, just as are the frequencies of various words. The individual probability of words and letters seems to have a correlation to their costs in time and effort when used; the total average cost in a message is generally minimized. The probability of occurrence for individual letters in the English alphabet is presented in Table 4:1.

Table 4:1 Individual probability of occurrence of letters in the English alphabet

Letter	Probability	Letter	Probability
E	0.13105	M	0.02536
T	0.10468	U	0.02459
A	0.08151	G	0.01994
O	0.07995	Y	0.01982
N	0.07098	P	0.01982
R	0.06882	W	0.01539
I	0.06345	B	0.01440
S	0.06101	V	0.00919
H	0.05259	K	0.00420
D	0.03788	X	0.00166
L	0.03389	Q	0.00121
F	0.02924	Z	0.00077
C	0.02758		

What about ordinary decimal numbers and their information content? The decimal notation with its repertoire of ten equally probable digits has an information content of $\log_2 10 = 3.32$, i.e. 3 bits (rounded, bits are always integers by definition). Therefore, in a message any sequence of digits may occur while sequences of English words occur according to certain rules.

Both ordinary letters and digits have a shorter notation and a more comprehensive information content than the binary system. The binary notation with its equally probable digits of 0 and 1 carries an information content of $\log_2 2 = 1$ bit and may of course be considered 3 times more excessive than the decimal notation and 5 times more than the alphabetic notation. But we may easily express every sign simply by the use of the binary digits 0 and 1 (the working principle of the computer!). Four digit binary combinations give 16 different words ($2^4 = 16$), six digit binary combinations give 64, and so on. Every time we add one binary digit to the word the possible combinations of new words are doubled. With the introduction of the extremely fast modern computer, the binary notation of the alphabet and decimal system has in a short time outdone communication systems such as Morse telegraphy and teletype.

Entropy and redundancy

The natural increasing of entropy in a closed system has its counterpart in the decreasing of information. The fact that information may be dissipated but not gained, is the information-theory interpretation of the second law of thermodynamics where entropy becomes the opposite of information. A message may spontaneously lose order during the communication, as occurs in bad telephone lines where a different kind of noise is present. Words in the conversation are lost and have to be reconstructed from the significant information of the context. From this point of view, information decrease is synonymous with entropy increase.

The fact that information may be lost by entropy but never gained is also seen in the act of translation between two languages. The translation never gains exactly the same meaning as the original. The translator always has to compromise between phrases that are more or less appropriate – in either case, some of the author's meaning is lost. Other sources of entropy are information input overload (see p. 81) and the importation of noise as banalities (e.g. nonsense entertainment and background sound effects). All this together will impair the organization and structure of a given message and thus culminate in a loss of meaning.

Earlier, entropy has been mentioned as the degree of disorder in physical systems. Transferred to information theory, the concept is used to inform us of the relation between a phenomenon and our information regarding it. More information is necessary to describe it when entropy increases. Consequently, what is arranged and structured needs less information to relate and comprises less entropy. A situation with many alternatives gives high entropy, while a decrease in the number of alternatives gives a lower entropy. Lower entropy is also the result when adding information to a system, something which was mentioned in the section on information-physics. The quantitative measure of entropy, interpreted statistically, therefore corresponds to the quantitative measure of uncertainty as defined in information theory.

Information-entropy has its own special interpretation and is defined as the degree of unexpectedness in a message. The more unexpected words or phrases, the higher the entropy. It may be calculated with the regular binary logarithm on the number of existing alternatives in a given repertoire. A repertoire of 16 alternatives therefore gives a **maximum entropy** of 4 bits. Maximum entropy presupposes that all probabilities are equal and independent of each other. **Minimum entropy** exists when only one possibility is expected to be chosen. When uncertainty, variety or entropy decreases it is thus reasonable to speak of a corresponding increase in information.

It is possible to calculate empirically the **relative entropy** of a certain language. An attempt to do this with the English language would need to begin with a study of its construction. Existing combinations of letters with their probabilities can be evaluated by using a dictionary as a starting point. If the word INFORMATION is used we may state that it has been chosen from a repertoire of 26 letters in 11 successive choices. With equiprobable letters every choice represents an entropy of 5 bits and the whole word yields $11 \times 5 = 55$ bits of entropy. The real entropy is however lower and is calculated according to the successive choices presented in Figure 4:7.

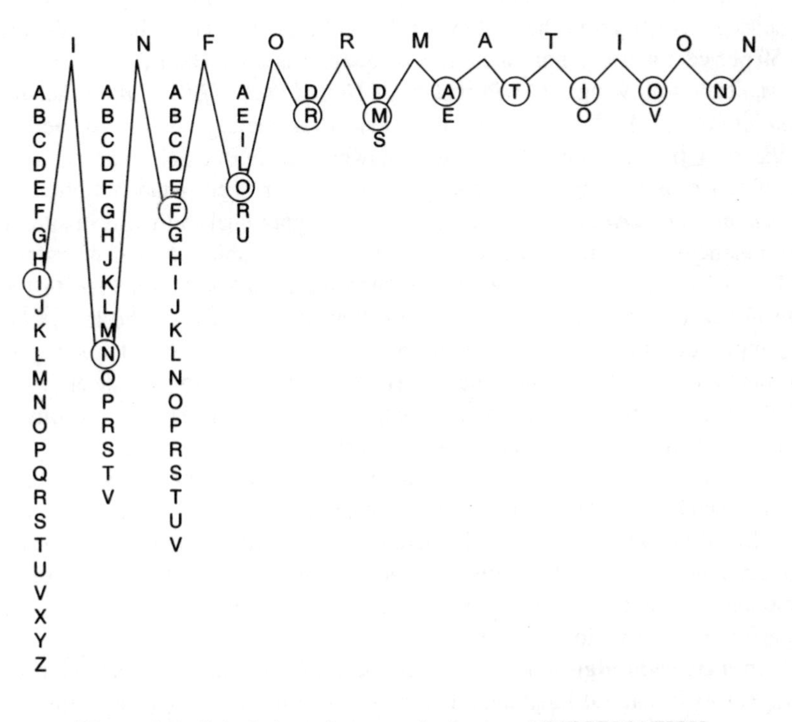

Figure 4:7 Calculation of entropy in the word INFORMATION.

The choice of the first letter is completely free and gives an entropy of 5 bits. The second choice is less free and must be made from among the 18 letters in column two; the English language does not permit any other combinations. This gives an approximate entropy of 4 bits and so also does the third choice. In the fourth choice the possibilities rapidly decrease and the freedom of choice of the information source is now practically reduced. The last seven letters are calculated in a similar manner but add very little to the information content. Therefore the acronym INFO is very often used instead of the full word.

The final calculation gives an actual entropy of 22 bits. A quotient between actual and maximum entropy gives 0.4 which is the value for the relative entropy in this case. Such a value may be interpreted as if the choice of the information source obeys 40 per cent free-will and 60 per cent compulsion according to the structure of the system. This component of an average message is what could be guessed owing to accepted statistical regularities inherent in our use of the alphabet and is called **syntactic redundancy**. It reflects the lack of randomness in our choice of signs or messages and denotes what initially seems to be superfluous, as we already know the structure of the system (the language).

Redundancy, the opposite of information entropy, is however both necessary and desirable in human language and is one of its most typical qualities. Empirical investigations have shown that a colloquial language has a redundancy of 50 per cent while a more technical language has less. The BBC newscast may transmit the following: '*The President of United States, Bill Clinton, has today announced....*'. For the majority of listeners the message is just as intelligible if the first part is omitted. There is however a substantial technical risk of interference; a hooting car may disturb the listener. The redundant phrase *The President of United States* is therefore wholly functional in the given situation. Redundancy is therefore *potential information*, available for us if necessary.

Normally we have no problem to interpret AB-ND-NCE -F IN-ORMA-I-N as *abundance of information* despite more than 25 per cent of the letters being missing. Our language may therefore be regarded as inefficient at first sight; an inefficiency which turns out to be highly necessary if it is to provide an inherent reliability. A language is always a compromise between basically inconsistent demands: precision and security in contrast to flexibility and efficiency. Our vocabulary is adapted to our everyday needs and we cannot have words for every special object or event. When everyday need demands important distinctions they do however exist; the Laplanders have eight different words for snow in their language while the British only have snow and sleet. In comparison with the decimal system, it is obvious that the alphabet contains a large amount of superfluous information.

Semantic redundancy is most easily described by the use of synonyms and paraphrases in natural language. The more extra names for the same thing, the greater the probability of making everything clear and avoiding misunderstanding.

Pragmatic redundancy is defined as the percentage of letters, words, etc. that can be removed from the receiver's message without changing his response. Total pragmatic redundancy in a message exists if a response intended by the sender has already occurred and is not repeated by the receiver.

The performance of a certain communication system must of course be designed according to the existing entropy of the information source. We must therefore realize the differences between human communication and machine communication. In communication between machines it is possible to reduce the redundancy in order to enhance the speed and efficiency. Technically, a complete reduction of redundancy in machine communication is only possible in a noiseless channel. Otherwise every individual error in a certain message would change the message into another one.

Channels, noise and coding

A communication channel is defined as any physical medium whereby we may transmit or receive information. It may be wire, cable, radio waves or a beam of light. The destination and/or source may be a person or a memory device like a tape or a computer disk. As a matter of physics, the capacity of a channel is defined by Hartley's law as $C = B \times T$. ***Ralph Hartley*** was a precursor to Shannon and his law from 1927 states that in order to transmit a specific message, a certain fixed product is required. The product (the channel capacity C) is defined by the bandwidth B multiplied by the time T. For the same message to be transmitted in half the time thus requires a double bandwidth.

Two main kinds of channel can be discerned: the *discrete* or *digital* channel applicable to discrete messages such as English text, and the *continuous* or *analogue* channel, long used to convey speech and music (telephone and LP records).

To apply information theory on continuous channels is possible by use of the *sampling theorem*. This theorem states that a continuous signal can be completely represented and reconstructed by sampling and quantifying made at regularly intervals. Sufficiently microscopic fragmentation of the waveform into a number of discrete equal parts bring about a smooth change from the continuous to the more easily calculated discrete channel. See Figure 4:8.

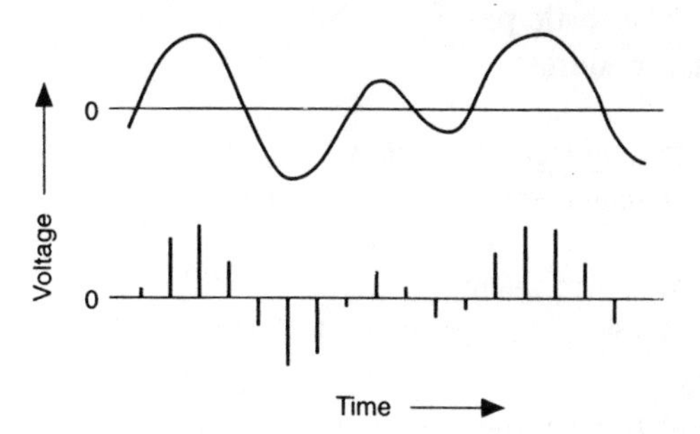

Figure 4:8 Measurement of amplitude samples in a continuous signal.

A message arriving at its destination may differ both structurally and functionally from the original because of **noise**. Thus, information is always lost to some extent, which means that the amount of information can never increase during the course of the transference of a certain message, only decrease. The existing noise in the channel is a kind of negative information. It is pre-empting the selective power of the channel leaving a residue for the desired signal.

Structural differences between a message sent and the message received are defined as **syntactic noise**. Examples are when a telephone call is distorted or when raindrops dissolve the ink script of a letter. Misinterpretation by ambiguity in the denotation of a message is called **semantic noise**. Bad spelling or choice of inappropriate phrases are typical examples. **Pragmatic noise** is not produced by the sender but is everything that appears in connection with the message and decreases its readability. A bad radio receiver may produce pragmatic noise.

However syntactically noisy a message is, it may not create pragmatic noise and can therefore at least theoretically be received correctly. A syntactically noiseless message may on the other hand fail its intention through environmental noise disturbing the receiver. Syntactic and semantic noise *may* generate pragmatic noise but need not necessarily do so.

Messages transmitted in electronic channels are always, to a greater or less extent, affected by certain harmful electrical influences or **transmission noise**. Of these types of noise, *white noise* is the most common. It is assumed to be random and non-Gaussian, making the message and the noise statistically independent. As its elements are independent of the communication channel it can be relatively easily filtered out by various technical means. System-dependent, or *black noise*, has certain dependencies between its elements. It is non-Gaussian, often highly nonlinear, and cannot be dealt with as separated from the system and the message. Up to now the complex nature of black noise has prevented the development of reliable methods for its calculation and neutralization.

The various categories of noise may be compared to the different kinds of uncertainties inherent in the use of natural language. According to ***Colin Cherry*** (1966) they are as follows:

1. Uncertainties of acoustic pattern
   (accents, tones, loudness)

2. Uncertainties of language and syntax
   (construction of sentences, use of synonyms)

3. Uncertainties of environment
   (disturbances by noise, background interference, etc.)

4. Uncertainties of recognition
   (past experience of the receiver, familiarity with the transmitter, etc.)

One method of solving problems caused by noise is to ensure a suitable amount of redundancy in the language used. General redundancy can also be achieved by different means such as repetition of the message, change of channel and form and the use of feedback in the communication process. Another method is efficient coding, which will decrease the negative effects of noise.

Information must always be transmitted in some physical form, something not to be confused with the form in which it is actually created. A **code** determines how a succession of symbols (numerals, letters, signs, etc.) can be replaced by another, not necessarily equal, long succession. The symbols used by the code may not be of the same kind as the original symbols used and all information, even the most complex, can be coded. Well-known codes are the Morse code used in telegraphy, the Baudot code used in teletype and the ASCII code used in the binary representation of computer languages. The transformation of a message by use of a code, *coding*, can however only be carried out for a meaningful message. Transformation of characters in a message apart from the meaning can only be effected by use of cryptography. The transformation of a message in order to regain its original content is called *decoding*.

The nature of a specific channel often requires coding because it is only in a coded form that the channel will permit the signal to propagate. Another reason for coding is that it will actually improve the operating efficiency of the communication system. Optimum utilization of the channel capacity is therefore a problem of matching a code to the channel in such a way as to maximize the transmission rate. A further, different reason for the use of codes is secrecy.

To study how a message can be most efficiently encoded (normally in an electronic channel) is one of the chief aims of information theory. Its subarea of coding is however both vast and complicated and only some of its main princ-iples may be outlined here. Let us however begin with coding of decimal numerals. Calculations earlier in this chapter showed that combinations of three binary digits were inadequate to express the existing numerals. Four binary combinations give 16 possibilities, a surplus of 6 (see below).

Binary number	**Decimal digit**	**Binary number**	**Decimal digit**
0000	0	1000	8
0001	1	1001	9
0010	2	1010	unused
0011	3	1011	unused
0100	4	1100	unused
0101	5	1101	unused
0110	6	1110	unused
0111	7	1111	unused

5 binary combinations give 32 possibilities, enough for the basic letters of the alphabet, but not for all the characters of a normal keyboard which requires about 50 keys. Here the use of 6 binary digits gives 64 possibilities, a surplus of 14 for a keyboard with no upper case. To express all the characters of a normal keyboard including upper case thus requires at least 7 binary digits, which gives 128 combinations.

If we want to express a three-digit number there are two different possibilities. One way is to let the information source select three times in the basic repertoire of ten digits which gives $3 \times 3.32 = 9.96$, i.e. 10 bits. The other way is a choice in the repertoire of existing three-digit numbers between 100 and 999 which gives $\log_2 900 = 9.4$, i.e. 9 bits. Obviously, the second method is a little more efficient than the first which in turn is more comfortable for human mental capacity. Our decimal system is derived from calculations with hands, feet and a limited short-term memory.

It is now possible to understand that *block coding* will reduce the number of digits used in the binary code when coding both numerals and letters. In a message, any sequence of decimal digits may occur, while letters in English words occur according to certain rules. To encode whole words by a sequence of binary digits is therefore more efficient compared to coding the letters individually. Here, inherent statistical laws of the English language give an average of 4.5 letters per word requiring 14 binary digits or 2.5 binary digits per character.

An epistemological problem related to message decoding is to detect if there is any message at all. This is especially relevant for radio astronomers listening for intelligent messages transmitted from extra-terrestrial sources. In such a one-way situation, the first difficulty is to identify the **metamessage.** A metamessage is an indication that there exists some kind of intelligent message. A sealed envelope or a floating bottle (if you are on the beach of an uninhabited island) are tangible examples of a probably existing content.

The radio astronomer has nothing tangible to expect, he must rely on the regularities of certain electronic patterns detected by himself or his computers. If something appears as a logic pattern, it must be regarded as relating to a time scale. A metamessage might very well be hidden in a time scale that exceeds the human life span.

If a metamessage is apprehended as such, attention is concentrated on the **external** message. Decoding of the external message requires knowledge of its implicit structure and symbol pattern. It is possible to add general instructions to the external message on how to decode it. The intergalactic gramophone records loaded on board the two *Voyager* spacecraft launched in 1977 have instructions on their covers on how to play them. The **internal** message is the real content of the transmission. Subtle ideas, emotions, and the possibility of 'reading between the lines' are typical of its content.

Applications of information theory

Information theory has its origin in telegraph and telephone engineering, which makes it natural that its benefits were first exploited by engineers in this field. These engineers thought of their devices as something which could exist in any one of a certain number of possible states and a message as something as chosen from a finite repertoire. Calculations of how much information the channel could carry per minute and how much was occupied by the specific message were done routinely. In this way it was meaningful to speak in terms of the informational effiency of, for example, a telephone channel and to compare it with rival coding systems. The indistinct and qualitative concept of information was here transformed into something precise and quantitative.

Of the many fields which benefit from information theory, we will mention only a few here. *Physiology*, especially interpreted in terms of GLS theory with its basic concepts of matter/energy and information, is strongly dependent on information theory. Calculations of how much information a special nerve fibre will carry per second, or how much information is processed per second in a certain part of the retina, are typical questions for information theory. The basic proposition is here that organisms are information systems.

Linguistics, particularly that sub-area engaged in synthetic speech research and the voice control of computer systems, rely heavily on information theory. The fact that physicists are massively involved with information theory has already been demonstrated in the section of this chapter on information physics. *Semiotics*, the theory of the phenomenon by which something acts as a sign to a living organism is quite naturally closely related to information theory. Also, *infology*, the science of presentation and reading of verbal and visual information is again naturally related to information theory.

The nearest application of information theory is, however, in *information science*. This area, with its strong interdisciplinary nature, is mainly concerned with the acquisition, storage, retrieval and general requirements of knowledge. As such it has also to examine problems of the information-rich and information-poor, information policies, copyrights and personal information integrity.

Review questions and problems

1. Try to give at least five different definitions of the concept of information.

2. What is the difference between a signal and a sign?

3. Linguistics distinguishes between three conceptual levels. Shannon's communication theory is relevant for only one of them. Which one?

4. What is the least possible number of questions to be posed to get the correct answer in the Game of Twenty Questions?

5. What will happen with a written message without redundancy if one of the letters becomes illegible?

6. Give an example how the concept of relative entropy can be used when analyzing the English language.

7. Show why block coding of a message is more efficient than individual coding of numerals and letters in the same message.

5 Some Theories of Brain and Mind

- **The need for consciousness**
- **A hierarchy of memory**
- **Brain models**
- **A model perspective**

'The brain is not an organ of thinking but an organ of survival like claws and fangs.' (*Szent-Gyorgi*)

In comparison with other animals, human beings are relatively slow and ineffective. Our physical strength and general performance fall far below that of animals of our own size. The human life-span is also short when compared with such different animals as the elephant or the Galápagos tortoise which limits the amount of human accumulated experience. We cannot remain in water for a long time, or fly; our body can only survive within a very limited internal and external temperature range and we must have oxygen to breathe.

We tire rather fast and spend half of our life time resting or sleeping. During the other half we are mostly hungry and engaged in eating or digesting food. The range and sensitivity of our sense organs are also limited; especially in comparison with birds and insects we are not able to see as far or to detect rapid movements. Our hearing and smell are inferior in comparison with that of our own dogs. All our senses are easily saturated by information - not too much per unit of time and of right kind, please!

Psychologically, we are very subjective beings, always looking for a meaning to our existence. We often compile fragmented data from our senses to construct an artificial whole, sometimes initiating very strange decisions and actions. We are excellent at adapting reality to our personal maps, remembering selectively and putting new facts into old moulds. Facts are generally interpreted to our own advantage. What is unknown frightens us and we attempt to neutralize it by creating myths, rituals and traditions.

The above description ranks human beings low on the scale of existing animals but, looking around, we find ourselves to be in some ways extremely successful. In spite of our shortcomings we somehow solve difficult problems and make reasonable decisions in critical situations. Man is apparently something more than a featherless biped (in the words of Plato) - thanks to his brain with its outstanding information processing, specialized in the weighing of uncertainty and making creative associations between different objects and error tolerances.

According to Miller's theory the brain is the equivalent of the decider and associator at the individual level. In Beer's *Brain of the Firm* the decider and associator at higher levels are treated as a metaphorical brain. In Lovelock's Gaia hypothesis we find humanity in the role of the huge, global brain of mother earth. Thus we uncover the idea of the brain function as a concept distributed among individuals in the higher levels of systems hierarchies. Genetically specialized individuals fulfilling the role of 'organizational brain' are not known in nature, although other essential functions such as the 'organizational reproducer' (queen bee) have been developed.

Apparently, the organizational brain is too important to be located in one single place as it manages functions directly involving the continuous existence of the organization. To minimize its vulnerability, a distribution strategy has been favoured by the development. Thus, in a sense Lovelock's 'global brain' exists everywhere and nowhere, something which also may be said of the individual mind.

When treating the human brain and its information processing the traditional *body/mind* problem is brought to the fore. This has been the source of the witticism from generations of lecturers: 'What is mind? No matter. What is matter? Never mind.' This problem has two aspects, one active and one passive. Usually the active aspect is known as the question of how the conscious mind by its will can influence the motion of material objects. The passive aspect questions how a material object such as the brain can evoke consciousness. More concretely the body/mind problem concerns free-will, intuition, creativity and the subjective unity of experience. The body/mind problem is today often interpreted in terms of quantum physics. The duality of body and mind reflects the basic duality between wave and particle – the origin of our physical existence according to quantum theory.

The need for consciousness

An important tendency of life on earth is that it resides in individuals separated from each other. Therefore, in the hierarchy of nature, consciousness is found at the level of organisms where highly centralized nervous systems with a brain exist. It is not possible, however, to imagine consciousness without an existing memory. Awareness of every kind had to relate certain of its constituents to earlier, not too short, reminiscences. In very simple terms, consciousness could be defined as the capacity of a system to respond to stimuli. The evolution of higher levels of consciousness with expression of will and decision-making appears to rest heavily on the pertinent accumulation of knowledge.

The benefit of a conscious mind has sometimes been questioned from a biological point of view and its existence has been interpreted as a secondary product of its own work with no special function *per se*. A different, but well-known point of view is the classic philosophical **panpsychism** which states that consciousness is like the force of gravity and the phenomenon of electricity, an inherent attribute of all matter, an omnipresent quality gradually manifesting itself when matter becomes alive. Consequently, human consciousness does not differ *in nature* from consciousness of elementary life forms or matter, only in degree and complexity.

Several quite reasonable perspectives nevertheless state that consciousness has a strong survival value. To define why it has that value is however very difficult if you compare human beings with early and primitive species such as ants and flies, which have survived for billions of years apparently without consciousness at all.

It seems, however, reasonable that consciousness has its most important role in handling entirely new situations where no prior references exist and difficult judgements have to be made. Consciousness allows for greater flexibility of behaviour than is achieved by preprogramming for even a wide range of possibilities. The precalculation of possible actions in a conscious mind implies no risks compared with real trial-and-error and increases chances for survival tremendously. To let bad ideas die instead of ourselves seems to be the very point of consciousness.

Maybe one can emphasize the special biological necessity of consciousness in connection with an all-embracing global catastrophe. Human beings may attempt to save their species below the water surface or orbiting in space until better circumstances return, thanks to a superior conscious intelligence and prognostic capability.

One of the most popular philosophical arguments for the need of a consciousness is the **antropic principle,** presented by ***John Barrow*** and ***Frank Tipler*** in 1986. This says that the laws of nature seem surprisingly well suited to the existence of life. The very nature of the universe seems to be creative and to include the existence of conscious beings, fulfilling the basic need for the universe to be aware of itself. It is in the end through human eyes that nature has attained the possibility to examine itself. The researcher Steven Weinberg has expressed it in the following way: 'The world is the way it is, at least in part, because otherwise there would be no one to ask why it is the way it is.'

Furthermore, without the existence of human beings reality would have neither form nor function. Consequently, the natural laws are designed to allow for the existence of conscious beings and reality exists as it does in order to create the proper conditions for human evolution. Our existence thus tells us something about the properties of the universe but also that the entire evolution of the universe is reflected in the human brain. As such the brain has got a built-in moral system to ensure its survival.

Regarding the antropic principle some followers prefer to interpret it in a more absolute way, thus coining the idea of a **strong antropic principle**. They proclaim that the existing state of the universe is inevitable and not a result of an accidental occurrence. It is impossible for us to consider another kind of universe as we could neither exist in it nor observe it (note the relationship with the interpretation of *superdeterminism* in Chapter 1). A corollary of the strong antropic principle is what has been called Tipler's 'beautiful postulate', namely that life, having once come into being, will continue for ever.

It is possible to impose a hierarchy of consciousness among living organisms from unconsciousness via consciousness to self-consciousness and omniconsciousness. In a broader context, these concepts are to be found in the higher **existential levels** which have been formulated by ***Ernst Schumacher*** (1978).

- The kingdom of minerals
- The kingdom of plants which compared with minerals have life
- The kingdom of animals which in addition have consciousness
- The kingdom of man which, in addition to minerals, life and consciousness, also possesses self-consciousness.

The possession of consciousness is positively correlated to an organism's **intelligence**. This concept can be defined as the capability to change a pattern of instinctive behaviour through the use of experience. To do this an aptitude to discern common elements of different situations is necessary, together with the ability to store this capacity for future use. Intelligence exists then among both animals and men. A real difference between animal and man is, however, very difficult to establish in spite of Schumacher's definition. Both kingdoms are strongly influenced by their mental world of **emotions** and **feelings** - the ultimate protective invention of nature to guard the organism. What the brain decides with its intelligence is thus motivated by feelings such as disgust or fear.

Some people state that the real difference between man and animal is embedded in human **morality**, a purely human capability to differentiate between good and evil. This argument is not very strong if one considers what people do to each other in everyday life; dolphins therefore seem to be absolutely superior in morality. To discuss the true location of morality may, however, be outside the scope of this book. Its origin and function may only be commented on as one of the significant qualities of the self-conscious mind.

A new variation of our ancestors, called *homo habilis*, the skilled man, emerged 2 million years ago. At that time the global climate suffered a dramatic deterioration with a series of ice ages. A main strategy for survival in an ever harder environment was social organization. This demanded a better awareness to cope with a quite new kind of complexity. The emerging morality became an intelligent adaptation to demands for an improved sensitivity: to know what was right or wrong in a new complex situation of agreements, coordination and mediation of knowledge. Ethical behaviour, therefore, is not a luxury but a necessity in social systems of high density and complexity.

Morality is founded on a capability for **empathy,** the comprehension of the inner world and feelings of other beings, the very prerequisite of civilizing (that is, sympathy is to feel *for* somebody, while empathy is to feel *as* somebody). The consequences of individual actions and their impact on other creatures may thereby be predicted reasonably well. What is good or destructive for others is thus related to what should be good or bad for oneself. Symbolic thinking expressed by a language and an explicit understanding of concepts such as me and you are here the prerequisite. The continuous training in handling this complicated world of symbols is believed to lie behind the development of the brain as well as all kinds of moral actions.

The quality of empathy has, however, an unavoidable complication in the possibility to exert intentional cruelty in order to reach one's goals. Moral awareness based on deep emotions gives man the sophisticated power of both destruction and healing in relation to his fellow beings. Without these emotions, no values working as guidelines for action, should exist.

Returning now to **consciousness**, it is here defined as the ability to create an inner mental world, an abstract model of the reality by use of memory - inseparably connected to perception. This predictive model is used on the external environment and is constantly redefined and tuned to the same with growing experience. Consciousness relates to basic feelings, e.g. pain, contentment, joy, and sorrow. Vertebrates are in general conscious beings, albeit to varying degrees.

Self-consciousness is then the creation of active models of a reality wherein the individual self is included; these models serve for both explanation and prediction. A self-conscious mind produces alternative models, even of a non-existing future. Thus both interpretation and anticipation of a future built upon various activities is possible. Self-consciousness gives a freedom of choice and a capability to manifest one's own will. Such an aptitude which offers the possibility to determine one's own fate contributes to a tremendously increased pace of development. While the human brain is the most significant location of self-consciousness, this is also recognized among the more developed animals, e.g. chimpanzees and dolphins. Emergent properties of self-consciousness are the use of languages and artefacts and time- and death-awareness among human beings.

Although dwelling in a world surrounded by physical things created by themselves, human beings mainly exist in the world of symbols. A breakdown in this world of symbols can lead to mental disorder, serious mental diseases, and is often in the background when suicide is committed. Thus mental diseases and suicide are unknown among other animals.

The highest mode of awareness, **omniconsciousness**, is based on a superior understanding of reality and exemplifies a new stage of development. It is characterized by some authors and philosophers as all-embracing and genuinely ethical and representing the ultimate degree of consciousness achieved by few human beings. Famous religious personages, e.g. Buddha, are said to have achieved this kind of consciousness. While this level is difficult to describe adequately using a lower-level language, it can be said to include unity with the environment without the loss of individuality.

The personal ego is understood to be part of an eternal, universal consciousness temporarily residing in the actual body. Persons approaching this level of consciousness see no reason to assert their ego. They are not dependent on the surroundings or its fluctuations and see their own misfortune, losses and criticism of self as real, but not crippling. To be in this state is to witness one's own actions as if watching someone else.

Apparently, the expansion of consciousness leads to an expansion into space. The possible merging of individual consciousness into one single mind stretching from person to person all over the world has always fascinated both philosophy and science. The resulting global reservoir of information produced by all mankind is often called the *Universal Mind*. That simple forms of consciousness merge to produce higher forms is part of the evolutionary paradigm (see ***R. Fivaz*** 1989). This idea is also integrated in both the Gaia hypothesis and the noòsphere of Teilhard de Chardin (see Chapter 3). The merging process itself has been the subject for several authors, among them the cyberpunk ***Rudy Rucker*** in his book *Software* from 1982.

When discussing concepts of consciousness, the **subconscious** and the **collective subconsciousness** have to be mentioned. These terms were introduced by the early psychoanalysts such as ***Carl Gustav Jung*** and ***Alfred Adler*** and imply that only certain parts of the mind can be embraced consciously. We can recognize our consciousness as part of a wholeness. Most often our consciousness is partial, a phenomenon we can appreciate after the interruption of a deep sleep with significant dreams. Collective subconsciousness is the realm of the **archetypes**, the inherited patterns for emotional and mental behaviour, shared by all human beings.

A very interesting and somewhat controversial theory regarding the origin of human self-consciousness has been introduced by the American psychologist, ***Julean Jaynes*** (1976). According to his **bicameral mind theory,** the birth of self-consciousness dates back 3000 years. Prior to that humanity had no concept of an ego or a personal space of mind within the individual. Of course there were social structures, a culture, languages and diverse experiences. But in terms of this discussion man was schizophrenic, instructed in every movement by insistent voices which were named as gods. Human action was ordered by the gods, not by free-will; feelings and decisions were the result of divine intervention.

From a psychological point of view, man was bicameral, with brain chambers corresponding to the left and right hemisphere. Non-linguistic activities emerging from the right chamber were transferred to the left by way of voices speaking within the head. Thus no individual decisions were taken and the responsibility belonged to the gods. According to Jaynes, the self-consciousness with its ego is a relatively new human quality, a product of historic evolution and therefore changing as time passes. Self-consciousness was not an essential for human survival; even today we perform the majority of our actions without being aware of them, e.g. driving the car or entering the subway automatically. In fact we are very often without consciousness of our ego without being aware of it. We cannot be aware of that which we are not aware.

Of course these ancient 'absent-minded' people were as we are, apart from the lack of a continuous stream of thoughts, regarding something else, which characterizes modern man. When something extraordinary happens modern man becomes attentive; historic man listened for the internal voice and instructions from the gods. These gods were no figment of the imagination, they were the evolutionary side-effect of a language capacity and the real will of the ancient man.

Cultural and political revolution, together with the rising importance of written language, gradually paved the way for human self-consciousness, which appeared in the form of a metaphor 'I' as a side-effect of a personal narrative, similar to that presented in the famous tales of the *Iliad* and the *Odyssey*.

New and dramatic insights into the nature of consciousness have been presented by the American neurophysiologist ***Benjamin Libet*** (1985). His starting point was the summation of the total information flow passing through our senses. The amount was imposing: the eyes alone transmit more than 10 million bits per second to the brain and all senses taken together more than 11 million bits. Experiments carried out by Libet demonstrated that a maximum of 40 bits per second was possible to perceive for a conscious mind and that the normal capacity was about 15 bits. In terms of communication, this channel capacity, or bandwidth is filtered down to 15 bits per second.

Such a tremendous reduction of the information flow entering the brain is analogous to that of switching from a floodlight to a spotlight in a theatre. You begin by seeing one or two faces on the stage, the spotlight roams about and something in the background becomes visible as you become aware. The conscious mind is extremely flexible but in each distinct moment it is limited to a rather specific area. We perceive much more than we are immediately aware of, but we are able to focus attention on whatever we want. Furthermore, this awareness concerns one sense at a time, e.g. the reflex to close our eyes when we want to hear better. This mechanism allows us to be aware of surrounding impulses without becoming confused by them – a kind of survival strategy.

Libet's main findings reveal the readiness potential of the brain. He has shown that the conscious will to carry out something appears half a second after its initiation by the brain. The awareness of the action is delayed and projected backwards in time, making us believe that our will preceded it. The consciousness is delayed and conceals this fact for itself by preserving an illusion of instantaneous awareness. This self-delusion is very practical when it is highly important to act instantly or instinctively. To react more slowly has to be done consciously. In both cases we maintain the impression of being in command of ourselves.

How then will the **free-will** relate to these findings? Hitherto it has been considered that the higher an organism climbs along the scale of evolution, the higher the degree of free-will and hence the ability to control and influence its own environment. According to Libet's experiments, awareness arises after the activity in the brain has started. But it will take 0.2 seconds from the conscious experience of the decision to its actual execution. Nonetheless, a conscious mind can stop the action before its execution, i.e. the consciousness cannot start the action but it may decide that it should not be executed. This **veto function** of the consciousness works through selection or choice as well as control of outcome of the will, rejecting suggestions presented by the non-consciousness.

The experienced feeling of self-command is the result of a sophisticated feedback of sensory data in time. We only cope with a small part of what we perceive – that part which gives meaning to the context. The delayed awareness enables us to present an adapted and coherent view of the world, an adjusted simulation which makes sense and which we are ready to perceive. Libet's work leaves little doubt that we are all at the mercy of influences of which we are very often unaware and over which we have virtually no conscious control.

A hierarchy of memory

A crucial step in the development of the human race was the possibility of using a language. It liberated the individual from the compulsion to merely learn from his own experience. We became a new and advanced kind of information processor. Use of natural language presupposes an extensive memory, something which in itself has been a basis for human survival and success.

Before the age of general abilities to read and write, the only place for memory storing and retrieval was in the brain. Memory was conserved through transfer from father to son and certain techniques to facilitate this process evolved. The storyteller, who had an important function as a living memory, used tricks like rhythm and rhyme to support the process of remembrance. Here we may see the beginning of poetry and literature.

Special memory techniques, **mnemonics**, were invented in prehistoric times and especially well-known among the ancient Greeks, who had formal courses in the 'art of remembering'. Central to every method was the organization of the material to be learned so that it could be retrieved when needed. Mnemonic systems are designed especially to impose meaning upon otherwise unrelated items. The range of techniques used includes the **method of places**, where some geographical location is used as a cue for retrieving items; the **method of associations**, where simple associations are sought between each of the items, connecting them into a meaningful story; and the **method of keywords**, where otherwise unrelated items are linked to numbers.

People with a special talent for remembering, who are said to have an **eidetic memory**, have always existed. They are noticed in the literature because of the problematic consequences of remembering literally everything that they come in contact with.

With the invention of memory-supporting artefacts and the art of writing, it became possible to store information and knowledge outside of the human brain. Human memory and intellect could be released from the task of remembering and be used in the creation and development of knowledge. In a sense, the invention of written language indicates the birth of science.

A more precise analysis of the many aspects of *information* will establish the nature of what really should be remembered and stored in the various kinds of memories. The following terms are used quite differently, depending on context and intention:

- capta
- data
- information
- knowledge
- wisdom

The arrangement can be seen as a continuum, whose parts lead from one to the next, each representing a step upward in human cognitive functioning.

If a basic event in the surrounding world is registered as a change in the state of a sensor (for instance a neuron) it makes sense to speak of *capta*. This change may be preserved for a certain period, being experienced for example as a lingering sense of heat. When some rules are applied to organize such basic representations of events, *data* is generated (singular *datum*). Numerals and the alphabet are such representations and the heat can be expressed on a Celsius scale. Naturally, data can be recorded and presented quite mechanically, without the perception of living beings.

Data reaching our senses and making us aware that something has changed or is going on is said to give us *information*. That is, we have cognitive or physical representation of data about which we are aware. In other words, we have been informed. Assigning meaning and understanding to information by the use of higher mental processes then makes it possible to speak about *knowledge*. This in turn may be transformed to *wisdom* when values are included in making judgements.

A hierarchy of memory, capable of storing relevant parts of the continuum or *intelligence spectrum* presented above, will have the following shape:

- Genetic memory, existing in the genes
- Immunity memory, existing in antivirus cells
- Accumulated experience and knowledge, stored in the brain
- Written information, stored on various materials
- Magnetic and optical information, stored in databases
- Encyclopedias, books, paintings etc., stored in libraries

- Metainformation, stored in universities, museums, gene banks, nature reserves, etc.

From this hierarchy it can be seen how different levels of living systems expand their memory capacity. While the cell stores its information in the genes and the organism in the brain, the group uses certain common instructions, calendars and almanacs. On the organizational level we find information stored in archives and accounts, whereas the national and supranational levels store their metainformation in valid laws and conventions. The main memory of the phenomenon of science must be considered to be the university. Culture must also be considered a memory wherein humankind has stored its increasing knowledge.

The memory expansion depicted in the hierarchy is only possible if a parallel development among memory artefacts take place. In reality it has always been an interaction between memory and its artefacts. Advanced artefacts make possible advanced information processing and *vice versa*. This interaction, made possible primarily through a capacity for language and writing, and the extension of images far beyond personal experience and lifetime, is a main driving force behind the exponential increase in the speed of human development.

The memory artefacts began with Sumerian cuneiform-inscribed clay and evolved towards pigment on papyrus, parchment and paper. Also stone and ropes were used to store information in the cultures of the Scandinavians and Incas. During the Middle Ages paper was the main storage medium, first as paper rolls and later as books. In the 19th century photographs and phonographs became available and in our own century film, shellac and vinyl records and a great variety of different magnetic and optical media have become available.

The main storage medium is nevertheless still paper. Our paper-bound cultural heritage has always faced sudden serious threats and we are now facing the risk of a collective memory loss through the decomposition of the paper. Before 1850 paper was a high-quality product and the raw material was taken from rags. After 1850 the wooden-fibre content increased – a raw material which now has begun to fall apart in an accelerating self-destructive process.

This problem may be compared to the destruction of Venice, which implies another collective memory loss. The buildings there now tend to slowly fall apart due to the diminishing ground-water level and the beautiful faces decompose due to acid rainfall. There are no natural countermeasures to these problems and the consequences are very difficult to forecast.

A third example of collective memory loss occurred in Alexandria. It is said that more than 1 million papyrus manuscripts in the large library were destroyed in AD 48 by the Roman Emperor Caesar. This is considered to have delayed European development by at least one hundred years.

Human evolution that was earlier governed by genes is today governed by ideas due to the accessibility of a huge collective memory, but also due to the extension of human sensing capabilities by artefacts. Examples of such artefacts are the telescope, the microscope, the telephone and so on. The principles behind knowledge accumulation and augmentation of mental capability presented here may be called **nōogenetics** (from the Greek *nōos*, meaning mind) to distinguish them from the ordinary biogenetic development. The nōogenetic equivalent to biogenetic mutations are new ideas, inventions and works of art.

Brain models

The brain has more than ten thousand million neurons, interconnected by means of a thousand times this number of synapses. Due to the speed with which biological membranes function this gives around 10^{16} interconnections per second. This is around one billion times faster than today's most powerful network computers. The most complex interconnective communication system in the world is the global telephone system – carrying only 10^{11} calls per year.

For some reason or other the human race emerged with this brain, oversized in performance compared to the needs of the bodily functions. This extra capacity was the basis for the extremely complicated nerve functions necessary for verbal communication. The richness of internal interconnections enables the owner of the brain to use symbols and therefore permits the development of a language. The number of neurons in the brain is thus an equivalent to the sum of stars in our galaxy. As a survival instrument adapted to the world surrounding us, the brain therefore probably is the most complex system which exists in the universe (with the exception of the universe itself?).

Today, a common belief among brain researchers is that the more the brain is used the better it will work. Intense use will cause the branches, called dendrites, in the neuron to grow. Dendrites are rootlike projections connecting the neurons. A typical neuron receives input signals from tens of thousands of other neurons. The more dendrites, the more interconnections promoting information transfer between different parts of the brain. Although divided into areas, each with a specific function, the brain processes information mainly in the same way in all of these areas.

Studies of the brain have shown that the length of dendrites may vary by as much as 40 per cent between different individuals. A most intriguing finding is that those who pursue intellectually demanding jobs have longer dendrites than those who do not. Two possible explanations for this phenomenon are: intellectually challenging life-styles cause dendrites to grow longer or having long dendrites leads people to live intellectually challenging lives. The first alternative, considered to be the most plausible, has received support through experiments with animals: rats raised in 'enriched' environments have been reported to show changes in brain structure.

A widespread attitude among researchers is that the brain is so complex that it will be impossible ever to embrace its whole function and capacity. A classic paradox formulated by the biologist ***Lyall Watson*** is: 'If the human brain was so simple that it was possible to understand its function, human beings would be so simple that they could not understand it.' It is a common view that the brain is a system which cannot be worn out; the brain grows with activity and will only be better by increased use. It is also assumed that we normally use only approximately one per cent of its total capacity.

The brain's storage capacity is literally astronomical and researchers believe that it can store every impression during a normal lifespan – with plenty of room left over. This statement that the brain can store every impression it has come in contact with applies to either of the two halves of the brain (or one of the two cooperating brains); like other essential mammalian organs, such as kidneys and lungs, it is duplicated. But unlike other doubled organs where each half of the pair works on equal terms, the brain pair is individually specialized with an established internal hierarchy.

The left brain seems to be specialized in serial information processing, while the right works primarily in parallel. Verbal processing and writing *via* letters, words, sentences, sections and pages is typically assigned to the sequential and analytical left side. Associative work seems to be assigned mainly to the synthesizing right side. Typically, this side processes more than seven elements in a very short time. It also recognizes musical patterns and chords and discriminates pitch as well. It discerns the form of the whole from its parts and recognizes complex visual forms (pattern recognition). Accordingly, two different kinds of information input may be treated simultaneously, compared and coordinated by both halves, thus creating a substantial and all-embracing impression. Functional differences between the left and right side can be listed as follows:

Left	***Right***
– Verbal	– Preverbal
– Analytic	– Synthetic
– Abstract	– Concrete
– Rational	– Emotional
– Temporal	– Spatial
– Digital	– Analogue
– Objective	– Subjective
– Active	– Passive
– Tense	– Relaxed
– Euphoric	– Depressed
– Male	– Female

There is also strong evidence that women have a better integration of the halves than men.

When the two halves of the brain become separated by accident or brain surgery (split-brain), the result is the emergence of two personalities, both with their own information sources and individual self-consciousness. Experiments with such persons involving screened-off vision show that an object held in one hand cannot be compared with a similar object in the other hand. There are simply no connections between the two eyes. When the artificial shield between the eyes is removed the two personalites are joined again to a single personality. The separated hemispheres have simply so many secondary interconnections via the brain-stem that their activities once again can be coordinated.

Within our Western culture an old tradition of analytical and rational thinking is coupled with a need for adequate expressions in speech and writing. The capacity for artistic work, intuitive thinking and creative fantasy is often seen as something less essential for both personal and societal development. From that point of view the left side dominates, sometimes creating the typical modern rational personality (sometimes with a touch of neurosis). For a harmonious development of the personality, society and its educational system must assign equal significance to the capacity of both left and right brain halves.

When discussing the overarching organization of the brain a rare quality existing in certain people must be mentioned. Known for more than 300 years, it is called *synaesthesia* and can be described as a multisensory integration in the experience of the surrounding world. Persons with this quality have a sensory crossover which makes them able to experience words, sounds, smells, sensations, etc., as coloured. The experience is involuntary and cannot be suppressed. Those with this faculty find their experience quite natural and can rarely understand that this mixing of senses does not occur in others.

Apparently, synaesthesia is a normal brain function in mankind but, for some reason or other, its working reaches conscious awareness only in a handful of people. It suggests that the brain has some kind of coordination centre where recall is reconstructed from numerous fragments of memories, stored separately, but accessed in an integrated way.

One of the best known concepts of the mind, called the **parallel distributed processing** (PDP) brain-model, is embraced by several neuro-scientists (see ***D. Rumelhart*** 1986). According to this model, intelligence emanates from the interaction of a great number of interconnected elementary units, the neurons. This slow and noisy apparatus performs real-time processing the only way possible: by working massively in parallel. Such a working mode means that a sequence, requiring millions of cycles if it were to be processed in a serial way, is done in a few cycles in a network of a hundred thousand highly interconnected processors.

In this model the brain work is regarded as a statistical process; no specially important areas are commanding the decision-making procedure. Decisions are made through cooperation between independent units, the neurons, creating reliability in turn through a huge statistical sample. Under these circumstances brain control is distributed, working in consensus but with no specific pre-calculated solutions. This kind of system is adaptive and flexible, constantly configuring itself to match the actual input. Although this process has neither classification nor generalization rules, it acts as though such rules were present. Learning itself results in a modified but more durable coupling density and a reconfiguration of the neural network, called **brain plasticity.** It is thus possible to say that the brain reprograms itself, creating thereby the very foundation for memory, learning and creative thinking. Supporters of the PDP model often say: 'There is no hardware and no software, there are only connections.'

Generally, human information processing is dynamic, interactive and self-organizing as well as superior in optimization and adaptation, a non-serial task. Also, it is robust and not too sensitive to inaccurate data; it handles incompleteness, ambiguity and false information very well. More specifically, in PDP-model terms the basis for these good qualities is that knowledge is globally stored in the existing network or structure and is continuously available. Essential information exists as **frames** or **schemata**, which are stored in flexible configurations offering the automatic supply of missing components, in a process of continual adaptation to meet the situation at hand.

Another key concept of the PDP model is **gentle degradation.** PDP does not know a critical amount of neurons when the network stops working. All parts may be seen as redundant and a damaged brain has a diminishing capacity corresponding only to the area of injury.

A special theory concerning selective mechanisms in the brain has been presented by ***Gerald Edelman*** (1987). It is called **neuronal group selection** (NSG) and is founded on the notion that many brain processes operate by natural selection, a kind of neural Darwinism. Also, processes governing brain formation and growth are of the NSG type. In the developing brain, specialization of cell function is determined by the characteristics of the other surrounding cells in which it finds itself. Cells of a developing nervous system tend to migrate to brain areas favourable for their further specialization. The selection unit is a number of neurons called the *neural group,* more or less specialized to respond to a certain pattern of input. Different groups may have the same input pattern but their reaction will differ a little according to internal structure and relation to other groups. Some groups are strongly specialized to react in a defined way to a certain input and are said to have *repertoires. Primary repertoires* become established shortly after birth and do not change. *Secondary repertoires* are established by change of connection strength between and among primary repertoires due to the situation at hand and are therefore in a constant flux.

Stable repertoires emerge by reorganization of old neural groups from other, less stable repertoires, and result in something new and more appropriate. In this reorganization a constant competition takes place between the groups resulting in a growing repertoire and stronger inter-group connections.

Besides the PDP brain model, several others are well known. Especially relevant here is the **triune concept of the brain** presented by ***Paul MacLean*** in 1972. According to this theory, the brain is organized in three main hierarchical layers, arranged with the oldest in the centre surrounded by the others, like the skins of an onion. The human brain is a complicated web of these layers superimposed on to each other according to the different evolutionary stages. It is reminiscent of a thousand-year-old town where old and new buildings exist side by side. Each of us thus carries the history of the whole biological evolution in our nervous system. The three main stages are presented in Figure 5:1.

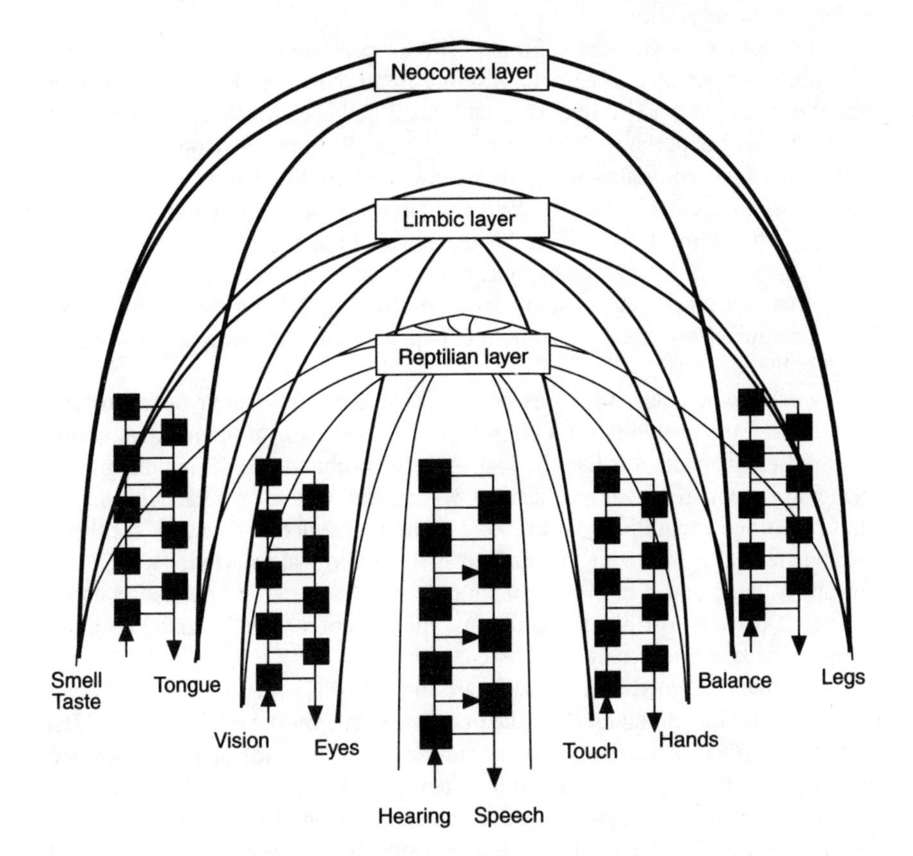

Figure 5:1 The triune concept of the brain.

The oldest layer, the instinct layer, is also called the **reptilian brain** and provides basic reflexes and instinctive responses. It can be characterized by aggression, rituality, territoriality and social hierarchy. The next layer, the emotional layer or the **limbic brain,** is the location of feelings and the important drive for altruistic behaviour such as the care of offspring. The third layer, the thinking layer or the **neocortex brain**, is capable of manipulating abstract symbols; it can analyze, associate, imagine and plan. Here is to be found the location of the essential human quality of intuition.

If the reptilian layer has a certain degree of consciousness, the limbic may be considered to be conscious and the neocortex wholly self-conscious. Unconsciousness and consciousness, old information and new thus exist side by side in a development similar to that of the city. The general function of these layers may be summarized thus: reptilian as biological, limbic as emotional, neocortex as intellectual. The coordination of all three layers defines what can be described as the human mentality.

Earlier we have stated that a prerequisite for consciousness is the existence of a memory. Apparently a memory residing in a conscious brain has no storage limitations. During a lifetime each individual gathers a tremendous amount of knowledge and experience for their own benefit. With the development of basic aural and visual communication, this knowledge could be transferred to other individuals to a certain extent. With the advent of a **language** among higher animals, communication was suddenly raised to a superior level and both knowledge and skill began an exponential accumulation. The nature of language implies an overwhelming communication desire as a guiding motive for human life. For humans an existence without communication with kinfolk is something unthinkable.

However, when self-conscious beings became aware of their own mortality they realized the problem of memory loss. The possessor of the memory sooner or later dies; its content is thereby lost. Another problem was the handling of the memory content: retrieval mechanisms were by no means on a par with the unlimited storage capability. An example of how knowledge could be conserved and transferred to following generations is the traditional storyteller who is both a memory bank and a conveyor of its content. To preserve knowledge and to transfer it between different generations, the ancient storyteller acted both as a memory bank and a conveyor of its content.

The above-presented brain models are mainly of a structural nature. One of the interesting functional models, the **brain resource model** introduced by ***Matti Bergström*** (1991) is based on a theory concerning the attitude of humans towards their the own brain. It is not possible to apply the same view on the brain as we do upon other human organs, such as the stomach, the liver, etc. The brain is the only part of the human organism which studies itself and this implies inevitable consequences for inherent values. These values cannot be excluded when studying the brain; without their application in human morality and ethics there are no qualities which may be called human. Furthermore, our values, coupled with feelings and fantasies, are necessary for the development of a healthy personality. The brain resource model does not discriminate between values and knowledge, between subject and object.

The growth of the brain is considered to be dependent upon an adaptation to an ever-more complex and demanding environment. A homogenous environment with no sudden changes permits survival with a small brain and an uncomplicated nervous system. The transformation of the planet, with shrinking oceans, formation of new land, severe weather changes, natural disasters, etc. demanded something more. Survival during these circumstances had to be built upon an improved capacity to receive, store and handle information, mirroring a more complex environmental structure. With this perspective the brain is simply seen as an **interface** between the environment and the internal world of the organism. Its purpose *vis-à-vis* the organism is similar to that of the skin: to adapt to and protect from the environment.

The brain-resource theory takes a very pragmatic view of the old mind/body problem. The mind embraces not only the abstract world belonging to areas like psychology, sociology or philosophy, but concrete physiological nerve processes as well. The 'wholeness' of the system is the mind and the different physical parts of the body. This view is exemplified by thermodynamic concepts: pressure, volume and temperature – macroscopic entities describing the wholeness of a system, but which are not possible to locate anywhere in a gas. The same goes for the mind, which it is not possible to pinpoint at a certain place in the brain.

In so far as the thermodynamic view has made possible both a demystification and a reasonable explanation of the complex concept of mind, it has been adopted into **neurodynamics**. Complex nerve systems with astronomical quantities of simultaneously propagating signals are not possible to analyze in terms of single pulses; the only way to understand such a mechanism is through a statistical approach.

One of the most significant qualities of the brain is its **value function.** This may be defined as the capability to choose and arrange knowledge according to an internal scale. Information and knowledge have no value of their own but are assigned a value as they become ordered and sorted. To assign values is to control and process information and to create negentropy, the opposite of entropy. Information overload, a serious problem in our current society, can never be handled with more information; a value should be designated to that which exists. To choose, evaluate and see a wholeness is the very core function of the human mind. A value-free science, for example in the nuclear area, is something of a paradox, when generating sophisticated knowledge to be used in an arms race and to promote chaos and potential destruction of the whole world.

A defective value function is always more critical than the quality of the adopted values. Children often reject a school culture overloaded with value-free knowledge which is so apparently without meaning for themselves; they turn instead to the hard gang morality and simple reward system of the street.

Another significant resource of the brain is its **creativity** function. The origin of creativity is found in the chaotic signal pattern of the brain stem, which acts as a random generator for new ideas if not restrained by social constraints. Some of these ideas may reach and influence the ordered levels of the brain, the neocortex, where a sudden change of mind occurs and is experienced. This mechanism may be functionally illustrated by catastrophe theory as a sudden and abrupt change when new ideas are born.

Our contemporary Western society does not support creativity. General social standardization, passivism and information pressure all too often restrain activity, spontaneity and the important work of the brain's random generator. The striving for a successful planning of our future also delimits creativity. To accept creativity is to lose the possibility of planning even the not-too-distant future. While creativity implies new and unpredictable knowledge which can lead to an unpredictable world, it also offers a strong survival value for adaptation to the future. Finally, creativity demonstrates how disorder and chaos are prerequisites of order and harmony.

A further consequence of the brain-stem as a chaos-generator is the general human fear of the unknown. The unknown internal ego and the unknown external unbounded world have slowly been mastered by the strategy of art and knowledge. Figure 5:2 shows how these concepts counteract the fear of the unknown in the human milieu.

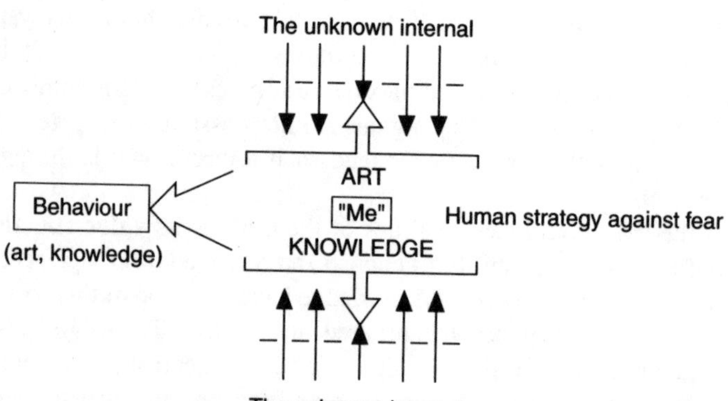

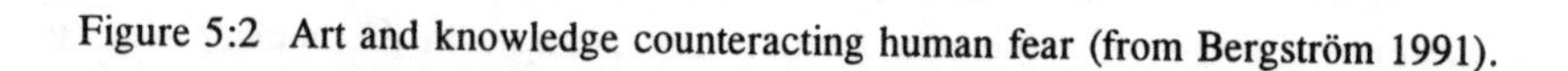
Figure 5:2 Art and knowledge counteracting human fear (from Bergström 1991).

Another key concept included in the theory is the **potentiality** of the brain. This term indicates that if one of the existing possibilities is realized by action, the others disintegrate. The responsibility inherent in each action is therefore tremendous; other worlds which are possible are destroyed when a choice is made. This dilemma has traditionally been solved in the Eastern countries introspectively, by decreasing external actions and increasing the internal potentiality of the brain. Could fewer actions instead of more be a more effective way to tackle this dilemma in the Western world as well?

The two competing states, the random and chaotic associated with the brain stem, and the fixed and ordered linked to the cortex, are the two bi-polar extremes of the brain. The simultaneous existence of chaos and order in complex, dynamic systems, however, tends to organize the content in a middle way in some kind of dissipative structure. It seems reasonable that such structures are generated in a self-organizing process of the brain, creating a new and relatively stable order. A large amount of dispersed information is suddenly joined to an integrated whole – a new idea, paradigm or method is born.

The **brain hologram metaphor** has been suggested by a number of scientists, including ***Carl Pribram*** (1969). A **hologram** (from the Greek *holos*, whole) is a kind of photographic image, created by illumination of a laser beam and has the following properties.

- The image is three-dimensional; it may be viewed from many aspects.
- A part of the hologram may be used to reproduce the whole image. The resolution of the whole image decreases as the area of the part decreases.
- Images can be superimposed and also individually recovered.

The holographic effect is the result of an interference pattern. To explain this phenomenon, let us take the analogy of three stones dropped in a pond of water. The resulting three circular wave-systems produce an interference pattern as shown in Figure 5:3.

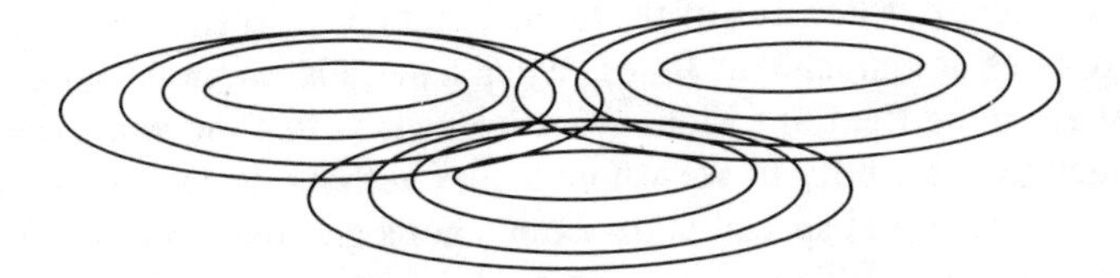

Figure 5:3 Interference pattern in a pond of water.

The pattern holds all information concerning the position of the dropped stones. A fragment of the pattern is sufficient to reconstruct the whole wave system (see Figure 5:4).

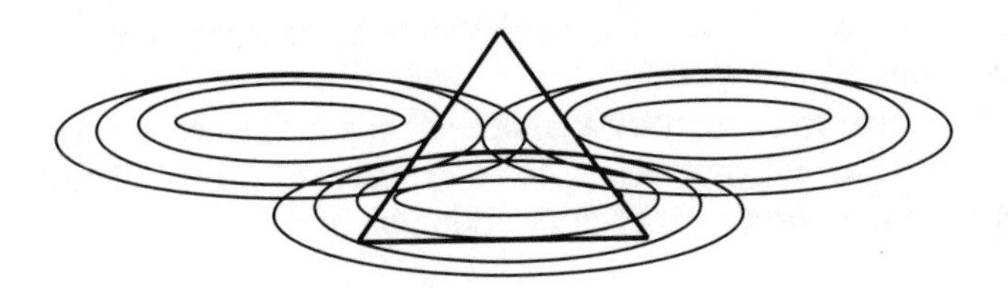

Figure 5:4 Pattern reconstruction using a fragment from Figure 5:3.

A holographic photograph is created by use of a laser light split into two beams. One beam is shines directly on the film while the other is reflected from the object to be photographed. Unlike ordinary photographs, the result is a blur covering the whole negative. This blur is a kind of interference pattern and when a laser beam is projected through the negative, the object reappears at a certain distance away from it. Different parts of the object are brought into focus by changing the viewing position (see Figure 5:5).

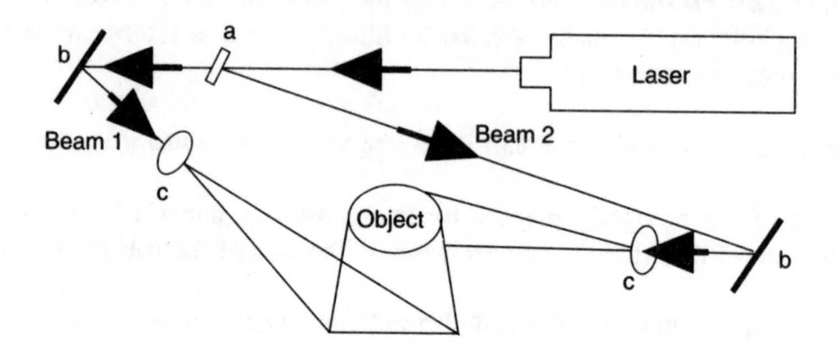

Figure 5:5 Holographic photography.

If images from a number of objects are stored, reflections from each item act as references for the others. Thus, one object may be used to recall another – a kind of **associative memory.**

An examination of the area of the human brain concerned with vision shows that it possesses holographic qualities. When a piece is cut away, nothing specific is lost from the field of vision. What happens is that the wholeness is seen less distinctly. The ability to see whole scenes without decomposing them into features, or to filter out special items from a homogeneous background are also holographic attributes. To discern a particular face in a crowd of people or to recognize a certain voice among all other vociferous voices in a cocktail party are typical examples.

A neural hologram may be imagined as a propagation of waves of dynamic neural activity. It should be obtained as a result of interference of neurons between two patterns. One sent directly acting on the near end of dendrites and one slightly delayed acting on the far end of dendrites. The input information is thus both distributed and redundant by the property of mutual convergence and divergence in neural pathways.

Of course, there is no sign of a complete correspondence between the real hologram and its neural equivalent. As an explanatory theory, though, it may shed light on certain processes of the working brain.

A model perspective

In the past many models of the mind have been constructed in an attempt to explain the phenomena of emotion, learning, perception and behaviour. None of them can however disregard the basic connection between the development of the hand and the brain. Freeing the hands by two-footedness made the brain able to grow. Without the use of hands there would have been no invention and manipulation of tools, no evolution of writing and consequently no information storage and tradition.

During the 19th century, brain models were dominated by analogies of mechanical, thermodynamic and even hydraulic processes. Most were dualistic, founded on the old Western thought that the body was a machine controlled by the brain, while the mind was separated from the body (and the brain). Between the world wars electric brain metaphors like the telephone exchange were predominant.

Since the 1950s, most models have been associated with the sequential computer with its central processing unit. The brain-mind distinction got its metaphor in the hardware-software distinction. Theories concerning the mind become materialistic by stating that physical and chemical properties of the brain, together with pertinent processes were enough to explain the mechanisms of mind. Thus a material phenomenon, the brain, created the immaterial mind.

In the 1990s, the perspective has changed again and the parallel computer, programmed as a neural network is the most current metaphor for the working brain. This metaphor is often discussed in cybernetic terms as the brain uses *both* positive and negative feedback in managing, for example, sensorimotor control. Such duality is uncommon in most ordinary control systems, due to the risk of instabilities, but is combined highly successfully in the brain. Positive feedback is used in a feedforward control course to predict what happens next, while negative feedback makes small corrections within the movement.

Today, most biologists seems to take a materialistic, reductionist view of the phenomenon of the mind. Its existence is the emergent properties of chemical and physical processes from a sufficiently complex control system in a living organism. The material base consists of neurons, interacting with other neurons to a certain extent randomly. This interaction creates information patterns existing in a context beyond ordinary time and space. Therefore, most researchers agree upon a distributed mind, occupying the total area of the brain where all elements are of equal importance.

The view of a distributed mind can be extended further, to the whole of the organism. The brain is part of a nerve system existing everywhere in the body. Thus consciousness exists ubiquitously, although on many different levels, throughout the whole organism, suggesting an extension of the distributed-mind perspective.

Review questions and problems

1. Discuss whether the traditional body/mind problem still can be considered relevant.

2. Can consciousness always be considered to have a strong survival value?

3. What is the relationship between superdeterminism and the strong antropic principle?

4. How does the problem of free-will relate to Benjamin Libet's findings?

5. Why can the invention of written language indicate the birth of science?

6. What is the difference between functional and structural brain models?

7. Use of the hands is considered to be a main factor in the development of the brain. Try to explain why.

Part 2: The Applications and How

6 Artificial Intelligence and Life

- **The Turing test**
- **Parallel processing and neural networks**
- **Expert systems**
- **Some other applications**
- **Artificial life**
- **Computer viruses**

'AI enthusiasts confuse information with knowledge, knowledge with wisdom and thinking with calculating.' (*Joseph Weizenbaum*)

Progressive scientific optimists in Western societies are often heard to promote two main projects: that of nuclear fusion and of artificial intelligence. While billions of dollars have been invested in both of these areas and very little has happened, proponents are still claiming that a breakthrough is very near. To understand why the difficulties seem to be insurmountable just before reaching the goal, it is necessary to investigate some of the main thoughts and attitudes of the artificial intelligence, or the AI, area. However, as we shall see, this investigation will be less about computers than about the nature of consciousness and mind.

AI has a very straightforward goal: to augment and improve human mental capability by replacing human mental activity with the activity of a computer. In order to do this, the construction of an **intelligence amplifier** is a consequence of the development in the modern technological society with its existing power potential and overall complexity. While our physical strength has been multiplied several hundred times by the use of conventional machines, no such comparable growth of human mental capacity has taken place. We may somehow amplify our inborn IQ sufficiently to give the overall comprehension that we so obviously lack in handling our dangerous strength. To do this, AI researchers state that it is necessary to find ways to formalize human cognitive processes in order to make them programmable.

Albeit a subarea in the discipline of computer science, AI has emerged as something more than an ordinary specialism. AI's success, if any, would have profound philosophical, ethical and practical implications for the entire human society. Also, deep feelings are disturbed when the superiority of the human race is challenged. Today, we do accept to be beaten in mental arithmetic and eventually in chess by our 'dumb' apparatus, but our intelligence, beliefs and feelings, constituting our total human pre-eminence, these we generally want to reserve for ourselves.

A closer look at the area of AI indicates that we can find some strong and influential supporters, the **functionalists,** who believe that within a few years computers will be capable of doing everything a human mind can do. For them, thinking is simply a matter of information processing and there is no significant distinction between the way human beings think and the way machines think (except that computers do it faster). When the program algorithms managing the computer's behaviour reach a certain critical point of complexity and function, intrinsic qualities such as free-will and consciousness will appear. Even human attributes, such as feeling pain, contentment and a sense of humour will emerge. Consequently, the statement that only brains within living individuals can become conscious is called 'carbon chauvinism' by the functionalists.

This school is also supported by several logicians and linguists who recognize that their specific area is in principle limited and therefore likely to be computable in the near future. In their view the mind is a machine operated on the basis of the known laws of physics and the human brain with all of its activities can be fully understood (by the human brain). Intelligence can be broken down into discrete modules with defined functions like perception, planning actions and executing actions. Electronic artefacts can then presumably be constructed to perform all these activities satisfactorily using programmed internal models.

Regarding the internal representation of the world, functionalists state that a richly articulated computer model poses no fundamental problems. The world is largely stable and can be sampled again and again by sensors. Relevant changes can be detected and added to the input when necessary.

A related faction includes the even stronger believers, the **behaviourists** who maintain that if a computer could be instructed to behave exactly like a conscious human being, then it would automatically assume the feelings of this creature. While we are still long way from this goal, the behaviourists claim that such mental qualities exist in today's computers. That is, every computer, even the simplest mechanical one, which performs a fundamental logic sequence of operations, has a low-level mental quality. The difference between low or advanced mentality and the existence or not of a mind is only a question of complexity, or of the number of states and functions involved. Thus behaviourists put an equals sign between doing and being when they state that to behave seemingly consciously is also to be conscious.

A conclusion which can be drawn from both the functionalist and the behaviourist views is that hardware is relatively unimportant. The software with its specific structure and algorithms is the critical part of the computer.

In contrast to these reductionist views of AI we have the **non-believers** who state that AI only exposes what genuine intelligence is *not*. They oppose the notion that the mind can be reduced to a machine operated on the basis of the well-known laws of physics. Existing knowledge is not enough to explain the mechanism of the mind and its emergent intelligence. Some of the non-believers claim that a true understanding of the brain is impossible because any explanatory device must possess a structure of higher degree of complexity than is possessed by the object to be explained. In other words, humans can never completely understand their own brains. In the eyes of many non-believers AI proponents show a mediaeval mentality in their attempt at an almost alchemical translation of dead machinery into a thinking being.

A well-known counter argument founded on the proposition that simulated or artificial intelligence should really be of the same kind as natural, has been propounded by the Berkeley philosopher ***John Searle***. It is a thought experiment called 'the Chinese Room' and supposes that a person sits in a closed room. This person cannot talk or write Chinese, nor does he understand anything of the language. His task is to receive Chinese sentences written on a paper through a slot in the wall. He then has to translate the text into English with the aid of an excellent dictionary containing exhaustive tables of Chinese ideographs. The translation is delivered though another slot in the wall to a receiving person outside of the room.

Although the translation is fairly good, the translator does not understand the *meaning* of the Chinese ideographs used, he only manipulates symbols according to a set of rules. As the action inside the room duplicates the function of a computer doing the same task, it is obvious that no real understanding, mental awareness or 'thinking' is present. The receiver has no idea of how the translation is arranged; so far as he is concerned, the whole process is a black box with a certain input and its corresponding output. In spite of this, the room arrangement shows every sign of having a very sophisticated, translating intelligence. It is therefore proper to say that the computer with its program serves as a *model* for human thought, not just a simulation.

Many non-believers hold AI enthusiasts responsible for clinging to the now outdated scientific belief in reductive analysis and mechanistic modelling and also for the belief that all the secrets of nature will one day be fully understood. They also state that AI researchers have forgotten their own starting point; to model intelligent behaviour, not to create intelligence.

Another argument against AI is that intelligence is defined in terms of living systems and is thus not applicable to non-living computers. Intelligence and knowledge are the results of biological functions of living systems with bodies, defined by autopoiesis, a quality not existing in computers. Living systems are capable of both self-replication, growth and repair. In the higher manifestations they have advanced nervous systems, part of which is the brain itself, managing individual physical existence.

Furthermore, living intelligent beings are biochemical creatures, guided by the very important capacity of their feelings. All emotions are inextricably tied up with a body and its states; electromagnetic machines controlled by a given number of lines of code do not have bodies. Memory retrieval in living creatures is also associated not with logical processes, but with emotional experience.

Therefore, attempts to get the computer to imitate the human cognitive system – to think for us – are in principle just as remarkable as to expect that the tools of a craftsman should do the job, not the craftsman himself. All this together makes AI programs end up in the same category as aircraft when compared with birds - imitations of a function but not of a process - not manipulating concepts but their physical correlations.

The definition of intelligence includes very essential social components; that a disembodied computer with no childhood, no cultural practice and no feelings should be defined as intelligent is both nonsense and a self-contradiction. The term artificial intelligence relates to a machine and has no relevance at all when making a comparison with human qualities. 'Are machines intelligent?' is therefore a completely irrelevant question of the same kind as 'Are machines resistant to AIDS?'

Regarding the human brain, it is the most advanced information processing system hitherto known; adapted as it is to the unlimited variety of life, it is far too complex to be treated as, or to be replaced by, any kind of human artefacts. The range of problems it has to handle includes the infinity of life and the infinity of human reflections. A computer can only equal or replace the human mind in limited applications which involve procedural thinking and data processing.

What really differentiates men from machines is the human ability to handle language - to comprehend any one of an infinite number of possible expressions is something that cannot be expressed in mechanical terms. Another significant difference is that human beings have the ability to diagnose and correct their own limitations in a way that has no parallels in machines. This power of self-transcendence implies the move to another level - to see the shortcomings of the system. The machine can only work within the system itself, according to rules which cannot be changed on its own accord. Therefore, being unable to break the rules set by software, computers cannot be considered creative, implying a further important difference in comparison with the human brain.

To be honest, a more realistic attitude seems now to be emerging among the new generation of AI researchers using the new breed of parallel computers. Intelligence is now to a lesser extent seen as a centralized, disembodied function, but rather as an epiphenomenon of the process *to-be-in-the-world*. Intelligence should thus be built through perceptual experience and learning, rather than by the implementation of an internal main model of the world.

Consequently, robotics has become an important area of interest where the hope is to trace intelligence through the experience of touch, sight, sound and smell. Using artificial sense organs robots should be able to build their own internal model of the world. The problem is however that both childhood and evolutionary history must be repeated and implemented if something called intelligence is to be replicated.

No matter what is the preferred perspective on AI, a dramatic point will be attained when transistor density in the central processing unit of a computer reaches the *human-brain equivalent*. This in quantity terms equals a number of about one hundred billion, that is, the amount of neurons in the human brain. Today, a common processor chip (for example, the Intel 486) comprises 1.2 million transistors. With present trends in chip manufacture it is not unrealistic to envisage a transistor density of one hundred million on a single chip within ten years. A comparable development within parallel computer processing should make possible the use of a thousand processors, thus realizing the brain equivalent with a possible clock speed of several hundred MHz. Which AI prophesies will then be realized, we can only wait and see.

The Turing test

Let us now return to the behaviouristic attitude and comment on the **Turing test.** This test was introduced in 1950 by the computer scientist ***Alan Turing*** (1912-54). Its purpose was to find out if a computer really could have the capability to think. The idea was as simple as it was brilliant. A person poses questions to an invisible respondent, either a computer or a human being. The impossibility of recognizing whether the answers come from a computer or a human being is said to be proof that the computer actually can think.

To ensure that the test is reasonably realistic, the communication must be exchanged in a technically neutral way, let us say with the help of a keyboard and a screen. All information about the situation is conveyed solely by way of the keyboard during the session. Also the human respondent must always tell the truth and the computer must lie if necessary to give the impression that it is a human. This point is especially important when the computer answers smart questions concerning, for example, number-crunching where it is known that the computer far exceeds the human being. (What is the square of 30497.034 times 2004.3 divided by 0.39794?)

The Turing test has inspired many efforts, among them the famous $100 000 Loebner Prize in Artificial Intelligence. The requirement for entrance into this contest is to submit a conversation computer program smart enough to mislead a jury of eight interrogators, to convince them that the conversation is taking place with a human instead of a computer. Hitherto, only a single-topic program has succeeded. *Whimsical Conversation*, written in 1991 by ***Joseph Weintraub***, was able to outwit four persons in the jury. In 1992 he won the bronze medal and US $2000 with the program *Men vs Women*.

While many natural reasons for the lack of success can be found, the main cause is what is called 'the common sense knowledge problem'. The problem is how to store in a computer and then access the total number of identified facts that human beings use in their everyday life. On closer examination, this task very soon takes on astronomical dimensions. To cope with a new environment, human beings base their actions on a recognition of certain similarities between the present situation and well-established past experience. Appropriate responses are gradually developed through trial-and-error, through training and imitation. Mentally, human beings advance from the past into the future, with the memory of the past going before them organizing the way new events are interpreted.

To store in a computer the myriad experiences of a lifetime - with small details such as how to give correct tip or to compliment a beautiful woman - is not technically possible. The fact that Eskimos have 40 different words for snow, Japanese have fifteen ways to say no and Arabs use 60 separate words for camel must be a nightmare for the AI programmer. To create a program that uses such details with flexibility, judgement and intuition then seems even more hopeless.

The use of language in the Turing test may put the computer at an unfair disadvantage. A special test, the Chess test, was therefore designed by the psychologist, ***William Hartston***, and a group of chess masters, to examine if it is possible to differentiate between (the intelligence of) man and computer in a game of chess. A set of chess positions is presented and an examiner compares the computer and its human adversary. The positions used were designed to show both the computer's and the human's playing strengths. Computers are very capable of playing complicated positions, instantly calculating the *immediate* consequences of possible moves. Human beings have their strength in recognizing the *long-term* strategic implications of a chess position.

In the test, one minute was given to each player, human and computer, to find the best move from eight presented positions. The results were measured according to a system which gave a negative score for responses belonging to the typical computer and a positive score for responses typical of a human player. The examiner had no problem in differentiating between man and computer because the latter always strove for a material gain and often captured pieces if possible. Even *Deep Thought*, the world's most capable contemporary chess computer, could not resist the material gain. The sacrificing of short-term interest for future benefit seems to be exceedingly difficult to implement in a chess-playing computer program. One can easily imagine what would happen if a slight change were introduced in the rules of the game in the middle of a contest. The human side would quickly adapt to the new circumstances, whereas the computer opponent would be left helpless.

In spite of several decades of hard work devoted to the development of AI, no evidence of what we normally define as mental qualities can be traced in computers in operation today. No computer algorithm, however complicated, has demonstrated some kind of genuine understanding and no computer has yet passed a serious Turing test. And of course no computer has shown the slightest hint of what is considered to define the human mind: emotion, free-will, creativity and ethical awareness. Nevertheless, from several other points of view, the work done by AI researchers must be considered to be fruitful.

Parallel processing and neural networks

Today a significant part of the development of artificial intelligence is carried out within the area of **neural nets** and **parallel processing.** Researchers and medical neuroscientists within this area are often referred to as **connectionists**. They hold that mental functions such as cognition and learning depend upon the way in which neurons interconnect and communicate in the brain. For a long time it has been known that the parallel processing in the brain is self-organizing. Several subprocesses which together deal with a major task are executed simultaneously in different parts of the brain's neural network. The main advantage here is the processing speed, completely superior to that of serial functioning.

The aim of constructing parallel computer networks consisting of artificial neurons is to approach something which functions in a manner similar to the real brain. These networks exist mostly as computer simulations, seldom as pieces of hardware. Experiments with neuron-like elements embodied in computers have been carried out for several years; this is now an established area on its own. The main advantage with computer neural nets is their capability to imitate (to a certain extent) brain plasticity. Thanks to this feature interconnections between neurons in the brain are able to change all the time to meet the task at hand. In short, the brain constantly reconfigures and adapts itself.

The main component of an artificial neural network, the neuron, is linked to its neighbours through adjustable connections. Like real nervous systems, which learn by adjusting the strength of their synaptic connections, the artificial neural network learns by adjusting the weighting on its connections. The strength of a signal transmitted through a certain connection depends on the weight of that connection. Most neurons or units of a neural network consist of three main categories: input, hidden and output. Signals are sent from a layer of input units to a layer of output units. On their way the signals pass through the hidden units, the function of which is to improve the computational power of the network. When any part of the input unit is activated, a pattern of activation is spread throughout the network. If the activation exceeds a certain threshold, each output unit sums up the arriving signals and switches itself on. See Figure 6:1.

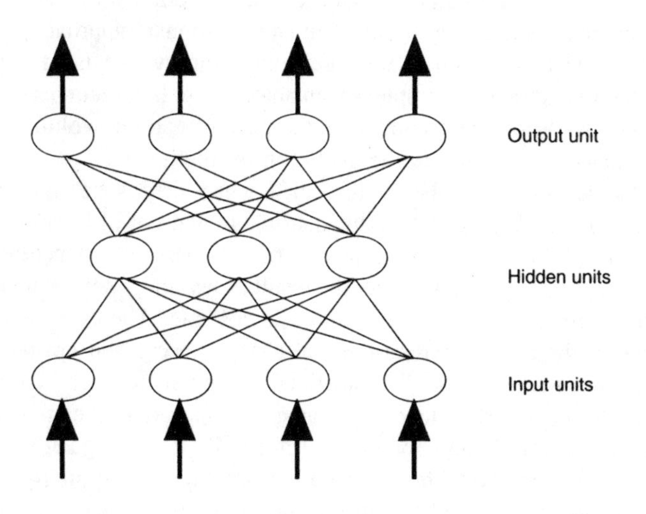

Figure 6:1 Components of an artificial neural network.

The network learns by comparing a certain programmed input pattern of activity with the resulting output pattern of activity, a process called **mapping**. More advanced modes of mapping occur in recurrent networks, where activation patterns emerging in the hidden units are recirculated through the network. For certain input patterns the hidden pattern generated is sent back to the input units with a small delay, coinciding with the next input. In this way the network remembers previous input patterns and learns of relationships between different input patterns. See Figure 6:2.

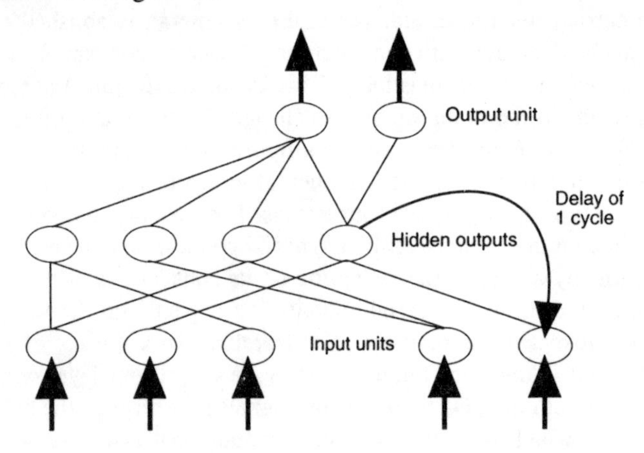

Figure 6:2 Principle of a recurrent network.

Neural networks work best when the input data may be fuzzy, since they do not depend on clear-cut yes/no decisions. Their decisions are made according to a complex averaging out of all the input they receive. Some proponents of neural networks see them as learning machines in an evolutionary approach to artificial intelligence. The assumption is that the underlying system is of a relatively simple structure and that its complexity emanates from large numbers of learned interconnections. With better computers the pace of natural evolution could be surpassed thousands of times in the evolution of intelligence.

Biologists do not subscribe to this idea when they state that intelligent behaviour is the result of a basic built-in structure and not of learning. A small insect with a few hundred neurons in its brain is extremely structured and its seemingly intelligent behaviour is not the result of learning. Regarding evolution at machine speed, this must be considered impossible. The evolutionary cycle must evolve at the same speed as the real changes occur and not the speed at which internal changes work. The artificial computer network cannot evolve faster than any other system that has to adapt to changes in the environment.

It must also be considered very doubtful if detailed knowledge of neurological mechanisms will ever reveal the true nature of intelligence. Similarly, exact and detailed facts about a computer chip will never reveal secrets of its associated software.

Expert systems

One of the beneficiaries of the spin-off effects of the AI area are expert systems, programs for problem solving in some scientific or technical domain. An expert system can be defined as a computer based system comprising the total sum of knowledge and rules of thumb which represent the highest level of expertise in a given and carefully limited field. Such fields may be medical diagnosis, mineral prospecting, deep-sea diving or operations of a similar kind which sometimes seek the solution of ill-structured, incomplete and probabilistic tasks.

Expert systems are sometimes very useful, but all too often we forget that they are fed with simplified and formalized average knowledge by ordinary human specialists. The fundamental characteristics of such systems are a lack of common sense and of a perspective of their own knowledge. That is, they do not have the capability to analyze their own experience. In order to exercise judgements an expert system employs certain intellectual rules of thumb, a phenomenon known as **heuristics**, sometimes described as the art of good guessing. Users of the system are thereby enabled to recognize promising problem-solving approaches and to make educated guesses.

An expert system consists of the following main parts (see Figure 6:3).

- A knowledge database containing domain facts, the heuristics associated with the problem and the understanding and solving of the same.

- An inference procedure, or control structure, for the choice of relevant knowledge when using the knowledge database in the solution of the problem.

- A working memory or global database, keeping track of the problem status, the input data for the actual problem and a history of what has been done.

- An interface, managing the interaction between man and machine and working in a natural and user-friendly way, preferably with natural speech and images. The interface must permit new data and rules to be implemented without change of the internal structure of the system.

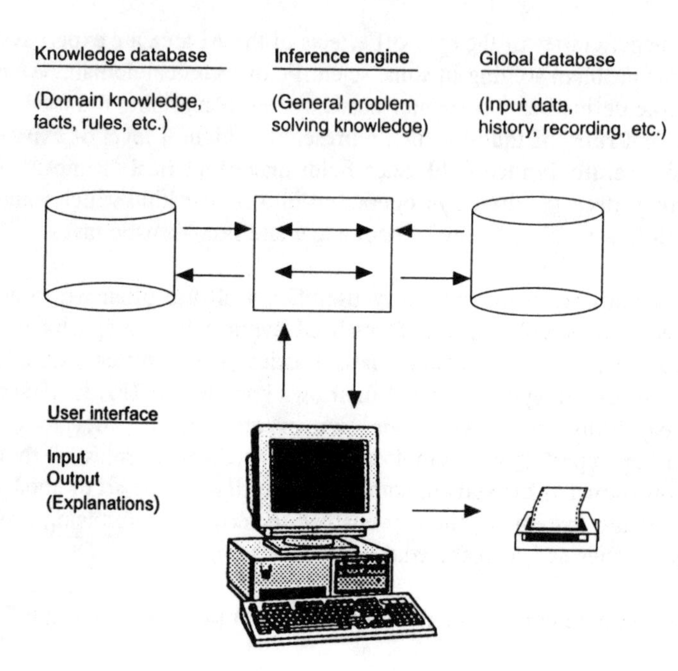

Figure 6:3 Expert systems architecture.

In medicine, diagnostic databases have been constructed which are claimed to be able to challenge the professional skill of an experienced medical specialist. Other well-known expert systems exist within the area of chess playing, where the capacity of the computer is now said to equal the capacity of a grand master.

A hierarchic model of skill acquisition has been proposed by, among others, ***Hubert*** and ***Stuart Dreyfus*** (1986). This model, which can be used to explain the deficiencies in the expert-system concept can be illustrated as follows:

1. Novice
2. Advanced beginner
3. Competent person
4. Proficient person
5. Expert

The skill of the novice is normally judged on the basis of how well he follows learned rules inasmuch as he lacks a coherent sense of the overall task. The rules employed allow for the accumulation of experience but soon have to be abandoned to make way for a further development. Problem solving is here entirely of an analytical nature and no special perspective is used.

The advanced beginner has gained a certain amount of experience and can perceive what is similar in previous examples. Actions and judgement may now refer to both new situational and context-free circumstances.

The competent person often registers an overwhelming amount of situational elements, all of which are to be taken into account. To manage the problem and find a solution, a hierarchical procedure of decision making is used. First a plan is established with which to organize the situation. Then a simplifying procedure is used to focus only on those factors important for the chosen plan. The remaining constellation of facts may generally be referred back to an earlier experience and reasonable conclusions can be drawn, decisions made or expectations investigated.

The proficient person, except for making decisions only in an analytical way, has the qualifications of an expert. Only when we reach the level of the genuine expert are real intuitive decisions possible.

The expert does not follow a general set of rules; instead it is a question of knowing when and how to break the rules and knowing what is relevant or not. If this intuitive, involved methodology has a distinctive feature, it is the very absence of a specific strategy. The expert now has a standard reference case but knows of numerous special cases.

A closer look at expert systems can give a more realistic understanding of their capacity and usefulness. None of the existing systems will be able to reach the level of a human expert; it is doubtful whether they will qualify as proficient. It is even less probable that the human ability to recognize, synthesize, judge and make use of intuition can be mimicked by a computer program. None of the semi-conscious processes in human thinking that emerge in those superior fuzzy concepts so typical of reality are to be found in expert systems. Human expertise is far too sophisticated to be formalized into a set of rules.

Expert systems can only present what has been stored into them. Often a craftsman makes things absolutely correctly without being able to explain why. Much of the existing experience and learning is what is called 'tacit knowledge', something which the hands, the judgement and the intuition bring about. Notwithstanding that such knowledge is systematically organized, it will have structures that cannot be expressed verbally. Of course, this is extremely difficult to formalize into a computer program.

It must also be kept in mind that even the highest ranked experts within a specific field have sometimes among themselves totally contrary opinions. When critical decisions have to be made, this problem is usually met by requiring the selected expert group to reach a consensus. The human capacity to merge individual irregularities into a collective agreement, a human qualified solution, is scarcely possible to implement in a computer. In other words, when a task is so narrowly defined that it can be performed with much less than the full human capacity of knowledge and judgement, then it is appropriate for an expert system.

In spite of the criticism presented here and, given that their limitations are understood, expert systems of course have their place as advanced intellectual tools. As it belongs to the lower levels of the skill-acquisition hierarchy, an expert system never needs thirty years of experience. It learns the main rules correctly the first time; it never needs practice and it never forgets. The precision and speed compensate for its blindness to situational elements. A good expert system performs about as well as competent human beings.

Some other applications

The potential application of both neural networks and parallel processing is most obvious in the areas of natural *language understanding* and *speech recognition*. To break the language barrier to knowledge, computer translation of scientific correspondence, papers and books would be a tremendous achievement. With the exponential proliferation of publications of all kinds, something which may be called the bulk barrier has to be overcome in all kinds of scientific work. The internal communication within the scientific community could be speeded up manyfold and the reinvention of the wheel occur less often.

Imagine the task of a scientific library which has the responsibility of translating, summarizing and storing a steady stream of papers in many foreign languages. To scan a paper, translate it into English, create an abstract, define keywords and store it in a database constitutes a well-defined undertaking for the area of artificial intelligence. A short step from this capability is the quality scanning of texts. The identification of grammatical and syntactical errors, stylistic shortcomings and a weak logical structure would utilize the whole power of AI. Another related task would be natural language interaction with databases. Today's communication demands some kind of formally rigid, abbreviated query language in the retrieval phase.

Speech recognition is an area which ultimately poses the hardest challenge for AI. One aim is voice recognition and translation of a telephone conversation into a second language, and there are already some applications for the input of unconstrained continuous speech using a diversified vocabulary, although with several weaknesses. The real task is to make it possible for say a Japanese and a Swede to converse in their respective native languages. No one has yet been able to create a program with enough grammatical rules to understand every sentence in a single language, let alone two.

A closer look at the **translating telephone** reveals that three technologies are required. First a device for automatic speech recognition, then a language translator and finally a speech synthesizer. Furthermore, the set of devices has to be duplicated; one for the hypothetical Swedish and one for the hypothetical Japanese side (see Figure 6:4).

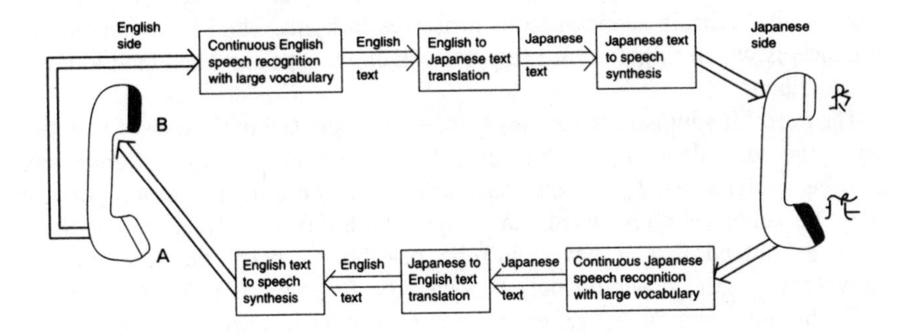

Figure 6:4 Main parts of a translating telephone.

The combination of the six modules should allow for a large, relatively unrestricted vocabulary, continuous speech input and speaker independence (little or no system training) in both directions. An interesting technical aspect relates to the intermediate link of text use. If an artificial language like **Esperanto** were used here instead of a natural language, the technical problems might appear less daunting. This is due to the fact that Esperanto is a very structured and simplified language, designed to be easy to handle although rich enough to facilitate good human communication.

Computers are however growing increasingly powerful and data storage is becoming cheaper. This fact was the impetus for an ongoing Japanese project to solve the translation problem by brute force. The JETS (Japanese-to-English Translation System) searches for entire sentences in a huge database consisting of all imaginable standard phrases with their variations. Because many similar sentence structures can be identified by a single string, the amount of memory needed is more restricted although still extensive. The main problem with this solution is however the prolonged search time.

Another application not to be forgotten is the *listening typewriter* which would be able to write down and edit the incoming speech on paper. A further use is *pattern recognition* which has applications in several areas: picture analysis, radar detection in military and civilian systems, cartography, etc.

Language understanding, translation and pattern recognition are extremely difficult areas for computer scientists and demand large computers capable of high-speed processing. The systems must be both self-learning and self-organizing in relation to the extremely comprehensive information to be worked upon. In spite of impressive funding and tremendous efforts in the last decade, any real breakthrough within AI has yet to come. One impediment must be the contemporary over-confidence in formal languages and the opinion that all phenomena can be expressed in logical concepts. Science seems to have been snared by expressions inherent to mathematics, statistics and logic. While the computer, working formally, that is, according to fixed models and algorithms fits this style, human beings act instead on the basis of direct communication and associations. Certain semiconscious processes in human thinking often emerge as concepts which might seem fuzzy but which are superior when reality has to be described.

The need for sophisticated AI is especially pronounced in the area of **robotics** where the equivalent of goal-directed behaviour found in biological organisms is to be implemented by mechanical machines. From a purely mental point of view, no robot yet constructed can compete with the capacity of an ant. Some sceptics contend that it never will be possible to reach the level of that marvellous insect. It should not, however, be forgotten that robots are often useful because they do *not* reflect the nature of human sensing and being.

In fact, robot technology does play an important role in scientific, economic and military affairs. The main development has however been within the area of practical industrial robots, where reliability and cost-effectiveness is more important than an advanced artificial intelligence. All over the world several manufacturing processes, both menial and unhealthy for human labour, for example, welding and spray painting, have been replaced by industrial robots.

The real need for AI becomes apparent in the special categories of robots used in bomb disarming, nuclear-environment repair operations and similar tasks. In contrast to symbolic problem solving as in language understanding, these special robots interact with the physical surroundings much as living beings would. They therefore have to be sensory-interactive and managed by a hierarchical control system. Problems to be solved may be illustrated by a robot manipulating one arm while pursuing an uncooperative target. Here the calculation of the coordinate's transformations have to be performed continuously, with delays of no more than milliseconds. A set of six equations with six unknowns has to be solved; this requires massive computational capability.

Artificial life

Artificial life, AL, is a fairly new science and a first attempt to define the area involves parallels with artificial intelligence. As AI relates to intelligence, so artificial life will relate to life itself. Like the AI researchers, investigators in artificial life are attempting to create this phenomenon in a medium other than the original and natural. A virtual medium wherein the essence of life may be abstracted from all details of its implementation is the computer.

The promotion of artificial life has many different stimuli. One of its gurus, Chris Langton, says: 'We would like to build models that are so lifelike that they cease to become models of life and become examples of life themselves.' A less spectacular motivation is to view AL as a generator of insight into the understanding of natural life. And, as always, when it comes to new and challenging areas, the old technological imperative lurks in the background: if it is possible to realize something, then do it.

The accumulated views on AL have crystallized into two main schools: weak artificial life and strong artificial life. The **weak artificial life** school wants to simulate the mechanisms of natural biology. The **strong artificial life** proponents strive for the creation of living creatures, a process sometimes called *computational ethology*.

Preoccupation with AL gives rise to several philosophical questions and also to some problematic definitions. A definition of life given in this book involves the use of autopoiesis. Suppose that a machine extracts energy from the environment, grows, reproduces and repairs an injury to its body. Is this machine living or not? Alternatively, is something missing in the definition? Complexity, self-organization, emergent behaviour and adaptive responses are attributes inherent to all living creatures and are attributes hitherto only occurring in the bodies of biological creatures.

Of all concepts to define life, complexity seems to have a key role. In living systems, the whole is always more than the sum of its parts. Living systems are tremendously complex and so many variables are at work that the overall behaviour may only be understood as the emergent consequences or the holism of all interactions.

Now, let us assume that all of the qualities listed above exist within a very advanced computer program. If human beings then watch on a monitor how artificial organisms grow, eventually mutate, reproduce and struggle for survival, is this life which is being observed and if so, does it reside within a body or not?

To solve this problem the **Duck Test** - a kind of Turing test - has been proposed by some AL researchers. This witty idea can be traced to the work of the French engineer, ***Jacques de Vaucanson***, who during the 18th century constructed an extremely realistic mechanical duck. 'If it looks like a duck and behaves like a duck, it belongs to the category of ducks.' In other words, if an artificial organism gives a perfect imitation of a living being and cheats an observer, it is living, no matter what it is constructed of. The core of life is then to be found in its logical organization, not in the material wherein it resides.

Most biologists agree in that the sole purpose of life is living, remaining and active; something which inevitably must bring about various kinds of self-interest. Therefore, to build a machine imitating life requires a construction oriented solely towards the maintenance of its own physical frame where life resides. Such a machine should not allow disconnection of its own electrical power and should react like the computer HAL in the film *A Space Odyssey - 2001*. Perhaps the HAL test would satisfy the needs of the computer ethologist.

AL seems to have started with something that appeared to be quite simple but developed into something very complicated - a complexity that was sometimes impossible to distinguish from what appears to be random. In 1968 the English mathematician ***Horton Conway*** working at the University of Cambridge, invented a self-developing computer game called the *Game of Life*. A deterministic set of rules served as the laws of physics, and the internal computer clock determined the elapse of time when the game started.

Designed as a kind of cellular automata, the screen was divided into cells whose states were determined by the states of their neighbours. The set of rules decided what happened when small squares inhabiting the cells were moved, thereby triggering a cascade of changes throughout the system. According to their moves, the squares could die (and disappear from the screen) or remain as survivors arranging themselves with their neighbours into certain configurations. New squares could also be born and placed on the screen. From Conway's description we read the following:

- Life occurs on a virtual checkerboard. The squares are called cells. They are in one of two states: alive or dead. Each cell has eight possible neighbours, the cells which touch its sides or its corners.

- If a cell on the checkerboard is alive, it will survive in the next time step (or generation) if there are either two or three neighbours also alive. It will die of overcrowding if there are more than three live neighbours, and it will die of exposure if there are fewer than two.

- If a cell on the checkerboard is dead, it will remain dead in the next generation unless exactly three of its eight neighbours are alive. In that case, the cell will be 'born' in the next generation.

What happened when the game started was that the game played itself, determined by the overall rules. Most of the simple initial configurations settled into stable patterns. Others settled into periodical configurations while some acquired very complex biographies. One of the most interesting discoveries was the 'glider', a five cell object that shifted its body with each generation, always moving in the same direction over the screen. It was an oscillating, moving system without physical mass, apparently some kind of evolving artificial life adapting itself and reproducing.

Later, Conway proved that the game was unpredictable; no one could determine whether its patterns were endlessly varying or repeating themselves. Embedded in its logical structure was a capacity to generate unlimited complexity; a complexity of the same kind as found in real biological organisms. The emergence of a stable pattern in the game is shown in Figure 6:5.

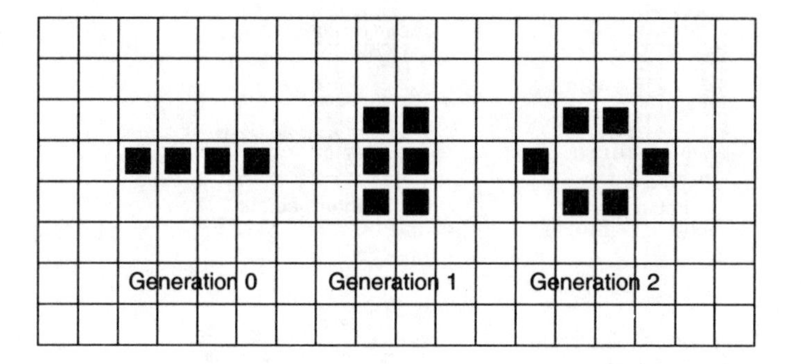

Figure 6:5 The emergence of a stable pattern in the *Game of Life*.

A common view among the AL researchers emerged that life in general was the result of a certain critical measure of complexity. When this level was reached, objects might self-reproduce quite open-endedly and thereby start a process of further complication. This emerging process was entirely self-organizing and apparently against all odds. But it happened over and over again as if life was inevitable, propagating over time.

For some researchers a corollary of this phenomenon was that the building material of the Universe was pure information. The ongoing creation and expansion of the Universe was an ultimate cosmic simulation and the computer was the Universe itself. Life's programming language was four-based instead of binary, built on the four basic chemicals creating the genetic code. If Nature seems unpredictable or random it has seldom to do with the lack of rules or algorithms. The point is our difficulty to see whether they are extremely complicated or overly simple. Therefore, nothing is done by Nature which cannot be done by a computer. Deterministic consequences of basic rules are however just as difficult to detect and understand in complex AL-simulations as in Nature.

A famous experiment in simulated biology was done by the American ***Chris Langton*** in Arizona in 1979. In his artificial world built on the computer screen, Langton was seeking the simplest configuration which could reproduce itself according to the same principles as living organisms that obey biological laws. His solution was a series of what he called loops, on the screen, resembling the letter Q and consisting of a square with a tail. The loops incorporated three layers of cells where the core layer contained the data necessary for reproduction. The rules dictating the generation flow allowed a certain growth of the tail until its decided maximum was reached when certain conditions were fulfilled. It thereafter turned 90° to the left three times until it had completed a new square. When the newly formed square resembled its parent, information was passed to the offspring and the two loops separated. The internal arrangement of the cells was then changed to an exact copy of the parent and the reproduction was completed (see Figure 6:6).

```
 22222222
2170140142
2022222202
272    212
212    212
202    212
272    212
21222222122222
207107107111112
 2222222222222

Initial
```

```
 22222222
2401111172
2122222202
202    212
242    272
212    202
202    212
272    27222222222212
21071071071071071112
 222222222222222222

Intermediate
```

```
 22222222  22222222
2017017012 2111170172
2722222272 2122222202
212    202        212
202    212        272
272    242        202
212    202        212
202    2122222222272
271111104104107107102
 2222222222222222222

Final
```

```
        2
       212
       272
       202
       212
 222222272      22222222
2111701702     2170140142
2122222212     2022222202
21     272     272    212
20     202     212    212
24     212     202    212
21     272     272    212
2022222202     21222222122222
2410710712     207107107111112
 22222222       2222222222222

Two new
```

Figure 6:6 Langton's self-reproducing loops.

An independent shape, consisting of pure information and reproducing itself, obeying deterministic rules of an artificial world, existed on the computer screen. Apparently, it was an organism consisting of patterns of computer instructions maintaining themselves through time, executing the very information of which they consisted. The process was the same as in real organisms and the question was whether it should be regarded as simulated or genuine life.

Several years have passed since the basic work of Conway and Langton. Similar work has been done by many other scientists and experiments with self-organizing, computer-insect colonies have shown how local interaction creates emergent, unprogrammed behaviour (see Langton 1989). Evolution of the same kind as in the natural world clearly exists in a virtual computer world. Symbiotic processes, natural selection including predation and growing stability and order, have been detected in simulations by ***Danny Hillis*** (1985) using his famous Connection Machine.

Computer viruses

Hitherto the only appreciable influence of AL on our own physical reality has been the consequences of computer viruses. Released computer viruses can be seen as an automated attack on a computer system with certain destructive aims. The result of the attack may have the following effects (Warman 1993):

- Interruption. Constant access to data stored in the computer is interrupted, sometimes for prolonged periods. This may have serious implications for business activities, for example.

- Interception. Access to internal information may be used to the creator's advantage.

- Modification. Internal information is changed by the attacker or replaced by misleading or inaccurate data.

- Fabrication. An extreme modification where data or transactions are entered into the information system in order to produce disinformation.

Computer viruses represent the claim for strong AL and the similarities between artificial and natural viruses are remarkable. Both are incomplete organisms fulfilling their sole aim of surviving by use of an host mechanism. By infecting, replicating and infecting again they preserve their individual code in different locations. However, the significant difference between a computer virus and a biological one is that the former has been written by a programmer for a certain purpose, whereas the latter arises spontaneously in nature.

Computer viruses have their own natural history and may be seen as a result of combination of predecessors like computer worms, logic bombs and Trojan horses. The **worm** is a piece of code which makes copies of itself intended to occupy accessible memory or disk storage space. When all space is occupied, the worm proceeds to the next free space into another computer via the network (if any).

The **Trojan horse** is a computer code which generally looks innocent and claims to be something other than what it is. When activated, the program begins its destructive activity, that is, deleting files from the computer's secondary memory. A certain kind of Trojan horse is the **trapdoor**, which consists of hidden extra code that enables the creator of the program to obtain easy and unauthorized access to the computer system without the normal log-on procedure.

The **logic bomb** is a relative of the Trojan horse with its program code remaining dormant until a specific circumstance will release it. The detonation may be initiated when a certain person is deleted from a payroll or by the system clock (a **time bomb**).

The various infecting and replicating strategies used by real viruses generally have their counterpart in the computer viruses. Classification of computer viruses according to their attacking strategy gives the following.

- Shell viruses. Create a shell around the code of the original program which therefore becomes a subroutine of the virus. This can be compared with biological viruses infecting cells and operating outside the cell nucleus.

- Additive viruses. Hook their code onto the host program. This can be compared with biological viruses which link their genetic code to the DNA in the cell.

- Exchange viruses. Replace the host code with their own code lines. This can be compared with biological viruses which replace the cell DNA with their own.

One of the most spectacular releases of a computer virus was made in November 1988 by Robert Morris Jr, a 21-year-old student at Cornell University. From the very beginning Morris lost control over his program. He could only watch powerless as it replicated itself through the university network and then beyond. This creation came to be known as the 'Internet Worm' when it congested all computers and finally shut down the whole network. When this was a fact some hours later, millions of users were affected. The losses were counted in hundreds of million dollars, including reprogramming and lost computer time.

Computer viruses have now existed for about 25 years, and hitherto have only evolved through the actions of human programmers. Their random variations have been clearly destructive for themselves and some kind of spontaneous evolution has not yet been detected. Effective vaccination programs written to neutralize all known computer viruses are now standard in most computer applications. In spite of this, many computer scientists and other researchers express concern for a future uncontrolled development. The possibility for a computer organism really going out of control by some kind of mutation, operating only for its own needs, may cause serious threats to our whole society.

Unengaged people see this attitude as overly pessimistic and state that one always turns the machines off if something seems to be dangerous. The counter-argument says that we never know when this point has come and that the turning off of our computers in itself is a threat to our high-tech society.

Review questions and problems

1. Why is the hardware considered relatively unimportant among proponents of AI?

2. What does the term *human brain equivalent* imply in the AI area?

3. Discuss if a computer ever should be able to pass a Turing test and if not, what should the main reason be.

4. Explain the functional principle of a translating telephone and interpret the use of Esperanto as an intermediary link.

5. AL researchers state that the core of life is to be found in its logical organization, not in the material wherein it resides. Discuss if this is a reasonable standpoint.

6. Considering the nature of AL and computer viruses, is it possible to imagine a completely harmless virus?

7. Does computer science know any self-mutating viruses?

7 Decision Making and Decision Aids

- **Some concepts and distinctions**
- **Basic decision aids**
- **Managerial problems and needs**
- **Four generations of computer support**
- **C³I systems**
- **Some psychological aspects of decision making**
- **The future of managerial decision support**

'What is the problem? What are the alternatives? Which alternative is best?' (*John Dewey*)

Systems can be described in terms of inputs transformed into outputs, as a process of fulfilment of a purpose, or the pursuit of a goal. The last is the equivalent of decision making and is a basic human activity. As a scientific area it is called **praxeology** and includes both normative and descriptive decision theory. In *normative* theory the goals are given and the decision method to reach them is prescribed and considered the ultimate. *Descriptive* theory is concerned with the way people actually make decisions without reference to the efficiency of their method. Belonging to the area of decision theory there is also *game theory*, a mathematical approach to problems of games involving conflict or competition among the participators. A decision which is not communicated and put into action is, however, no decision at all. Therefore decision making and communication must be considered as totally interdependent.

The inner core of decision making is to make things happen or to prevent them from happening; something which includes both prediction and control. In hierarchic, multilevel systems such as organizations and societies, this kind of operation is mostly dedicated to a specialized group, the management or the government. Mechanisms of decisions are strictly related both to the system to be managed and to the nature of human information processing.

Like other intellectual activities, decision making can be learned by the use of sequential steps. Its complete success can, however, never be guaranteed because it involves components of both uncertainty and creativity. To make choice is part of mankind's condition. As an uncertain and erring creature, he sometimes makes decisions, the results of which are often both unknown and unintentional.

Since its arrival, the computer has been used as a means for decision making, by augmenting the human capacity to gather, store, retrieve, and process various kinds of data. As the key component of an *information system*, it has greatly enhanced the transformation of data into information. As sub-component of a *knowledge system* (of which an information system is only a part), it has facilitated both understanding and the accumulation of wisdom.

Today, both air force and naval officers are trained for decision making in simulator installations which resemble the cockpit and the bridge. Decision making by management staff can also be practised in computerized business simulators that are made to resemble the ordinary decision environment as closely as possible. Beer's 'decision room' or *Frontesterion* is an example of such an environment (see Beer 1979).

In the light of the close interconnections within the modern world system, bad decisions may have both directly devastating consequences and unforeseen effects. The understanding of decision-making processes, particularly in association with computers, has therefore been a high-priority area of systems science. Knowledge of human capacity, specialty, and weakness – alone or together with computers – is a basic requisite for every decision maker. Systems theory and related areas such as **information theory**, **computer science** and **management cybernetics** have long been devoted to the study of decision making. A common assumption of these areas is that all organisms are information systems.

Some concepts and distinctions of the area

In order to understand the nature of decision making (and the need for computer support) a description of the characteristics of a decision situation must first be given. It can be defined with the items below.

- A problem exists
- At least two alternatives for action remain
- Knowledge exists of the objective and its relationship to the problem
- The consequences of the decision can be estimated and sometimes quantified

Generally, different *aspects* such as economic, environmental, and political fitness create the background of a decision. In this setting the decision maker first has to find existing *alternatives* and then make a *choice* between a set of them (normally small). In turn, every alternative has *consequences*, connected to the aspects via the alternatives. Given these sets of aspects and consequences the decision maker has to choose the best alternative.

All decision situations belong to one of the following three classes.

- Decisions under *strict uncertainty*. Here the decision-maker is unable to know anything about the situation. Quantification of the uncertainty is not possible.

- Decisions under *certainty*. In this case the decision-maker has full knowledge of the situation, and the consequences of the decision can be predicted. The alternative which has a value not less than the value of any other alternative is chosen.

- Decision with *risk*. In this situation the decision-maker is able to quantify the uncertainty by assigning probabilities, generally known in advance, to each alternative. Note that the level of risk increases exponentially as more data are left out.

Furthermore, a problem can be **solved** or **dissolved**. A solution occurs when one of the alternatives in the existing repertoire is chosen and then implemented. If the whole view of what is defined to be a problem is changed, or if a completely new alternative is *generated* to handle the situation, the problem is said to be dissolved.

For the decision-maker, the ultimate goal is considered to be to arrive at as effective and precise decisions as possible. But in reality, different qualities of a decision must be accepted as inevitable. ***R. Ackoff*** (1970) has categorized the quality of a decision into two levels. **Optimizing** implies finding the best possible existing solution. The tools available in such situations, apart from the decision-maker's own intuition and experience, are different types of models to support the decision-maker. The majority of these models are, however, either of a mathematical or statistical origin and of a rather complex nature. With the aid of a computer these models can be handled more effectively and efficiently, often by use of an algorithm. An *algorithm* is a step-by-step procedure (often of a mathematical nature) that guarantees that an optimum solution is achieved.

Satisficing is attaining a certain minimum quality level for the decision, enough to solve the problem but not necessarily more. The satisfier seldom evaluates the existing alternatives, because the first acceptable solution is considered to be as good as all the others. To satisfy is to use the principle of least effort. Most satisficing problem-solving strategies are based on *heuristics* - rules of thumb that are good enough for most decisions.

Interesting arguments questioning the need for optimization have been published by ***C. Holling*** (1977). He notes that ecological systems strive to maximize options rather than to delimit them by selection of an optimal alternative. From a human point of view, the possibility of a bad choice and pertinent failure is not rejected. Instead a strategy which minimizes the cost of such a choice is applied. In that way efficiency is sacrificed for adaptability.

A theory of decision making has long existed in economics, being associated with the idea of *homo economicus*, the strictly rational decision maker. This ideal human being has the following qualities:

1. He can always make a decision if faced with a number of alternatives.

2. He ranks the consequences on a scale of preferred results (a value-scale).

3. His order of preference is always transitive (first A then B, not C then A).

4. The first alternative is always chosen (utility-maximizing).

5. The same choice is always made if the situation is repeated.

According to ***Herbert Simon*** (1976) the process of *rational* decision making is an act of choosing among alternatives which have been assigned different valuations. It involves the following process:

1. Listing all of the alternative strategies.
2. Determining all the consequences that follow upon each of these strategies.
3. Comparatively evaluating these sets of consequences.

Simon, however, admits that total rationality is an unattainable idealization in real decision making – who can be aware of all existing alternatives?

The task of making decisions can generally be seen as an (iterative) procedure of information gathering and processing, summed up by the following keywords.

- **Intelligence** Find raw data to be processed

- **Design** Evaluate different alternatives of action

- **Choice** Choose one of the alternatives

- **Implement** Firmly establish the chosen alternative

- **Control** Check that orders are obeyed and make necessary adjustments

The decision-maker can only obtain information and thus only have real knowledge about future development within a rather short time frame. This frame is defined by the decision-maker's and his assistant's specific knowledge about the expected and agreed consequences of already known projects. The first step in the above process will inevitably suffer from an inherent discrepancy called 'the information dilemma' or even the 'uncertainty-relation of decision making'. It is associated to the need for explicit and actual information and states: 'the precise information is not timely, and the timely information is not precise'.

It is important that the whole procedure is understood to be cyclical; repetition and feedback of certain steps are virtually always indispensable. Even *re-definition* of the original problem and the existing alternatives may be necessary.

Using the above keywords, the following more detailed steps in the decision-making process can be elaborated.

1. Identify the problem (recognize a situation that requires decision/action)
2. Gather the facts which will affect the decision
3. Generate possible alternative solutions
4. Specify the alternatives
5. Select the best one
6. Gain acceptance by motivating/explaining the basis of the decision to other members of the decision-making group
7. Communicate the decision to all those affected
8. Put the decision into action
9. Supervise the execution
10. Follow up the results

The decision-maker thus has to choose the best option given the existing set of consequences. Note that a *non-decision* exists as an ever-present alternative (unfortunately most often the worst!). A classic reminder to the decision-maker is ever present: 'You may be so preoccupied with doing things right that you forget to do the right things.'

When ready to make his choice, the decision-maker will meet four basic types of difficulty:

- How to compare the alternatives with regard to different aspects of the decision.

- How to compare the alternatives within each aspect.

- How to estimate the probability that the given consequence will occur if a certain action is taken.

- How to estimate the value of the consequences.

A short look at the internal nature of problems which have to be solved by different kinds of decision reveals that they may be structured, unstructured, or semi-structured. **Structured problems** are those for which we can define an explicit procedure to solve the problem. An example is the construction of a schedule for the use of existing classrooms in a school. To solve **unstructured problems** the decision-maker must show judgement, evaluating capacity, and insight into the problem-definition. Political decisions are often unstructured as their success is dependent upon the changing opinions and hidden beliefs of the people. **Semi-structured** problems are partly structured and partly unstructured.

If the structure of a problem is related to the *operational*, *tactical,* or *strategical* decision levels identified in most major organizations, the examples in Table 7:1 will be typical.

Table 7:1 Decision levels and problem structures

Kind of problem	Operational level	Tactical level	Strategical level
Structured	Construction of timetables	Purchase of service-cars	Location of new railway
Semi-structured	Stocktrading	Dimensioning PR budget	Choice of new business office
Unstructured	Composing of cafeteria menu	Hiring new managers	Choice of new company logo

Not apparent from the table is the general tendency to find the majority of structured problems in the operational level. Also, most semi-structured problems are to be found at the tactical level, while unstructured problems are most common at the strategical level.

Structured decisions seldom involve managers and can hence be made by lower level personnel or by a computer. Semi-structured decisions are by their nature appropriate for managers with computer support. Computational complexity, problem size, and the precision of the solution often make strictly managerial judgement insufficient. Unstructured problems are not able to be formalized in a technical sense and hence are impossible to feed into a computer. The nature of the problem, the volume of data, or the lack of an appropriate method make any decision entirely dependent upon human experience and intuition.

Basic decision aids

During the past years a number of mainly mathematical techniques have been developed to assist the decision-maker. These techniques are called *decision aids* and their use is intended to maximize the probability that the chosen decision is the best one. Among the better known are the following:

- decision trees
- decision matrices
- linear programming
- game theory
- linear regression
- mathematical modelling
- forecasting.
- PERT (program evaluation review technique)
- critical path method

Of these, decision trees and decision matrices must be considered genuinely basic and are the only ones treated here. Decision trees and decision matrices are not dependent on advanced mathematics or the use of computers.

A *decision tree* is a model which gives a visual presentation of the structure of a decision situation. It has branches spreading out from nodes like a tree. The nodes are of two types: decision nodes (usually presented as a small square) and event nodes (usually presented as a small circle). From each decision node all the potential decisions branch out. Seen as series of decisions, the second step is dependent upon the first step, the third depends upon the second, etc. If risk or uncertainty is associated with each step, these qualities are gradually accumulated. An example of a decision tree is shown in Figure 7:1. Note the probability assignment given for each branch.

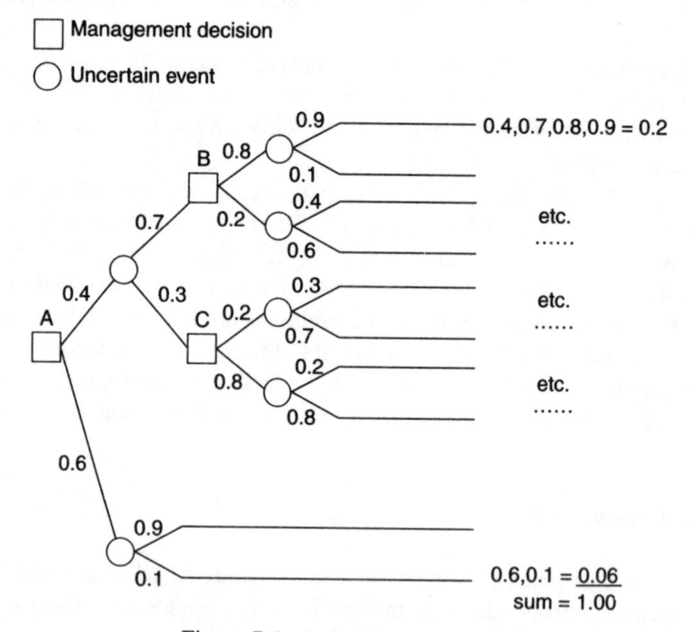

Figure 7:1 A decision tree.

To make a decision involves in practice cuting off a branch of the tree that it is no longer possible to reach. A transition takes place from openness to closure, something which presupposes a moment in time when all necessary information is collected.

A *decision matrix* is another aid for enhancing the choice of the best alternative when the various options have been sufficiently identified. The value of a decision matrix increases when the number of alternatives increases. The use of the matrix can be clarified by explaining four consecutive user steps. *Step one* is to identify all the alternatives which seem reasonable in the pertinent situation and to assign them to the matrix.

In the second step criteria are established to provide a basis for the selection of one alternative over another. Here each quality is given a number in relation to its order of importance. The numbers are summed, and weighting factors are assigned by dividing this total into each item's individual number. An example with eight criteria is given in Table 7:2.

Table 7:2 Criteria ranking and weighting factors

Criteria	Ranking factors	Weighting factors
Quality A	1	8/36 = 0.22
Quality B	2	7/36 = 0.19
Quality C	3	6/36 = 0.17
Quality D	4	5/36 = 0.14
Quality E	5	4/36 = 0.11
Quality F	6	3/36 = 0.08
Quality G	7	2/36 = 0.06
Quality H	8	1/36 = 0.03
	Sum = 36	36/36 = 1.00

The third step is to assign rating factor values to all the present alternatives considering the selection criteria. A scale from 1 to 10 is applied, where 10 is best. The assignment of these values is best done one criterion at a time, evaluating each alternative according to that particular criterion.

The fourth and final step in completion of the matrix is to multiply each ranking factor by its corresponding weighting factor and to record the product to the right in the column. Ultimately, all products are summed as in the completed matrix of Figure 7:2.

	Selection criteria								
	Eight different qualities, labelled from A to H								
	A	B	C	D	E	F	G	H	Sum
Weight	**0.22**	**0.19**	**0.17**	**0.14**	**0.11**	**0.08**	**0.06**	**0.03**	1.0
1st Alt.	**3/0.66**	**6/1.14**	**1/0.17**	**9/1.26**	**8/0.88**	**9/0.72**	**2/0.12**	**3/0.09**	5.04
2nd Alt.	**6/1.32**	**9/1.72**	**9/1.53**	**10/1.4**	**9/0.99**	**2/0.16**	**5/0.3**	**7/0.21**	7.62
3rd Alt.	**10/2.2**	**7/1.33**	**8/1.36**	**9/1.26**	**7/0.77**	**6/0.48**	**8/0.48**	**8/0.24**	**8.12**
4th Alt	**2/0.44**	**9/1.71**	**3/0.51**	**10/1.14**	**4/0.44**	**3/0.24**	**9/0.54**	**9/0.27**	5.55
5th Alt.	**4/0.88**	**2/0.38**	**7/1.19**	**9/1.26**	**2/0.22**	**2/0.16**	**2/0.12**	**9/0.27**	4.48

Figure 7:2 A completed decision matrix.

Mathematically, the highest sum can be 10, since all weighting factors together must be 1 and the highest ranking factor value is 10. In this example the third alternative had the highest value and was thus chosen. The great advantage with this matrix is that all steps in the decision process are clearly documented and the idea behind a given choice can easily be demonstrated.

Besides the techniques described here as decision aids (used inside or outside a computer), the computer itself is a kind of decision aid if sensibly used. The general principles for use of this type of assistance is presented in Table 7:3.

Table 7:3 Basic principles for general computer aid in decision making

	The computer is employed to present:		
Human information used for	**Data**	**Consequence calculations**	**Decision advice**
Inspiration	Improved environmental	What-if calculations	Advice-giving ideas
Idea control	Computational basis	Filter human ideas	A choice of better alternatives
Decision	Databanks	Alternative testing	Accepted/ rejected alternative

The horizontal axis shows an increasing degree of computer processing. The least processed output is data which can be further processed together with human supplements. By this means what-if questions can be asked and sensitivity analysis can be performed. Of course, the computer may be able to recommend a certain course of action if enough information and structure has been programmed into it.

The vertical axis shows how output from the computer can be used by man with increased degree of human processing. On the lowest level the computer is only used as a source of inspiration while at the highest it is used as a decision adviser.

The table can also be used to describe how the good decision-maker can interact with his assistants. The computer, however, has a tremendous advantage in that it has no ambition for a career of its own. It can never be a yes-man.

Managerial problems and needs

Today, all kinds of organization are constantly exposed to changes in a dynamic and moving environment. In order to survive they have to predict these changes and be prepared to respond quickly and adequately. When it comes to commercial enterprises the following major driving forces seem to be the most influential today:

- Development of new production technology
- A shifting competitive balance, from domestic to global competition
- A slower expanding market
- Changing market preferences with a demand for high quality at a low cost

With this background it is natural that market share must be captured from competitors in a global marketplace. Success is possible if the enterprise gets the information about the changes as they occur and possesses the qualifications to process it adequately. The competitive edge belongs to those enterprises which can rationalize and use the new technology in planning, operating, directing, and control of available resources. Use of decision support systems of various kinds has thus been one of the critical success factors in modern enterprises. It is here defined as '*a computer-based information system designed to support decision-makers at any level, working with semistructured or unstructured problems*'.

With respect to this definition it is important to understand that it is not a question of how to man a computer system. Instead it is a question of how to equip a small group of decision-makers with individually adapted computer tools. The benefits of implementing such a system have been investigated by several researchers (e.g. Kroenke 1989). These are:

- Increase in the number of alternatives examined
- Better understanding of the business
- Fast response to unexpected situations
- Ability to carry out *ad hoc* analysis
- New insights and learning
- Improved communication
- Improved control
- Cost savings
- Better decisions
- More effective teamwork
- Time savings
- Making better use of data resources

Some typical tasks facilitated by a decision support system have been pointed out by Bidgoli (1989) and are:

- **What-if analysis** The effect of a change in one variable can easily be measured in relation to others. If labour cost increases by 4 per cent, what is going to happen to the final cost of a unit? If the advertising budget increases by 2 per cent, what is the impact on total sales?

- **Goal seeking** Goal seeking is the reverse of what-if analysis. How much must be sold of a particular unit in order to generate an increase in profit of 5 per cent?

- **Sensitivity analysis** Using sensitivity analysis will enable the detection of the most influential and critical variables of a calculation. What is the maximum price you can pay for raw material and still make a profit? How much over-time can you pay and still be cost-effective?

- **Exception analysis** This calculation monitors the performance of the variables that are outside of a predefined range. It highlights the region that generated the highest total sales or the production centre that spent more than the predefined budget.

- **Trend analysis** Before making a prognosis the past has to be studied. Useful building blocks of the past are time-series which are computer processed and current long-term development or trends. These, in turn, are extrapolated into predictions of the future.

- **Revenue generation** Can be used to assess sales strategies, effectiveness of the sales force, allocation of sales personnel to territories, and appropriateness of the commission plan. As for marketing strategies, effectiveness of PR and advertising can be studied. Order-processing effectiveness can be assessed by examining the time required to fill orders, the number of backorders, and so on.

- **Purchasing** Can be analyzed by comparing the cost and terms of goods pur-chased to industrial averages. Cash flow requirements for future periods based on past experience can be estimated as well as the payment policy (when to pay invoices, whether to take discounts and so on).

- **Personnel and payroll** Plans can be developed, and changes in them can be carried out. Lead time for hiring and training employees on the basis of sales plans and manufacturing can be calculated and acceleration or retardation of the plans simulated.

- **Asset control** The estimation and calculation of the market value of different types of assets can be facilitated. The impact of changes in asset depreciation schedules and the evaluation of asset control effectiveness can be performed by estimating losses due to theft, accidental destruction, or bureaucratic bungling.

- **Product planning and budgeting** Can be performed. The cost and schedule of developing new products can be estimated and the financial result of planned product portfolios can be forecast. Trial budgets can be developed and analyzed in order to estimate their impact upon different types of divisions, etc.

- **Manufacturing planning** For example, how large production facilities need to be in order to meet the manufacturing schedule, is conveniently calculated on the decision support system. Monitoring of the manufacturing process and assessing the costs and benefits of different alternatives is also possible. Quality measurement and subsequent detection of the causes of changed quality is another possibility. Finally, the effectiveness of production scheduling can be studied and better ways of organizing the production effort simulated.

Four generations of computer support

During the 1960s, when the first generation of useful all-purpose computers reached the market, a kind of decision support called *electronic data processing*, **EDP**, became available. It was designed to do the task of implementing decisions which had already been taken. Very large quantities of data were processed, each in accordance with a well defined and programmed series of choices and conditional branches. In a loose sense this processing can be considered to have comprised decisions, although the 'decisions' were nothing more than mechanical recognition of patterns without the aid of human judgement. Its designers were typical computer scientists and it was applied in transaction systems, among other uses.

Later the concepts of *Management Information Systems*, **MIS**, emerged. These systems were the result of a cooperation between computer science and the area of management. The underlying need grew out of an increased internationalization and competition among organizations in a more complex world. Visions of computers revolutionizing the business sector with on-line, real-time systems supporting rational, quantitative decisions by large amount of data were common. In reality the MIS has become a tool for routine middle management decisions by supporting simple bookkeeping, updating inventory and calculating cost over time, and so on.

As a predecessor to MIS, the *Decision Support System*, **DSS**, arrived in the 1970s as an instrument primarily for modelling. Also, this system was intended to be a tool for decision-makers, but became a little too complicated for the average executive. DSS systems were generally geared for special, limited processes of decisions involving multidimensional models, *ad hoc* analyzing-reporting, and consolidation. Today, these systems are used mainly by controllers and analysts for calculating, budgeting, simulation, and aggregating.

During the 1980s two generations of *Management Support Systems*, **MSS**, were introduced. The first generation comprised mainframe computer systems with links to personal computers. A private database was included in the system, which had a menu-driven, character-based user interface. The system-support was manual, usually administered by a special computer department. The software consisted of different software packages for different functions.

At the end of the decade a change of generation took place among management computer systems and a second generation MSS emerged with the client/server technology. No particular databases were used and communication took place with databases belonging to other, external, agencies. The user-interface became a graphics screen used for data-driven models, consolidations, and reports. The software packages were integrated for all functions.

At the beginning of the 1990s the demands on computer systems grew rapidly. From the executive's viewpoint the data provided was produced too slowly and was not good material for calculations. With the rapid technical development that led to powerful workstations, relational databases, and improved data communication, new demands arose. These demands were met by the concept of the *executive information system*, **EIS**, which emerged in response to the demands. (Some computer scientists use the acronym **ESS**, *executive support system* which is synonymous with EIS.)

The idea was that EIS should be a support for top-level executives in their daily work involving decision making, planning, and controlling. Information should therefore be presented, processed, and handled in a very simple manner appropriate for the layman. The main point of the system should be flexibility and user-friendliness.

A closer look at EIS shows that the system is customized for a particular decison-maker and it is used directly by that decision-maker without an intermediary. The system is extremely user-friendly, requires relatively little learning time and gives the user support for all kinds of problem definitions. It is designed to process large amounts of data from both internal and external sources and gives the user the possibility to predict and simulate various courses of action.

EIS may be described as a computer supported presentation/analysis tool which collects data from other systems or from its own database. It facilitates *drill-down*, that is, the selection of information and navigation from an aggregated to a more detailed level, in a well-structured way. Requested data is processed by the user according to his wishes. Its main functions lie in analysis and modelling where information from different sources is combined. Information can also be *distributed* by the system to employees in other departments by electronic mail. The presentation of information is achieved via high-resolution graphics on screens and by tables and text.

Today, decision support is also found in *expert systems*, **ES**, which can adapt their own rules in a manner predetermined by another set of rules (see p. 191). By the use of artificial intelligence and by imitating processes that human experts unconsciously perform, the system can serve as an advanced problem solver and decision maker. Note, however, that the problem to which an ES is applied must be complicated enough to require an expert. In principle, the expert system can deliver its own solution at decision time and do the whole job itself, although a responsible decision maker is unlikely to give up his right to decide.

Figure 7:3 shows how the various kinds of computerized decision support systems have been used at different management levels.

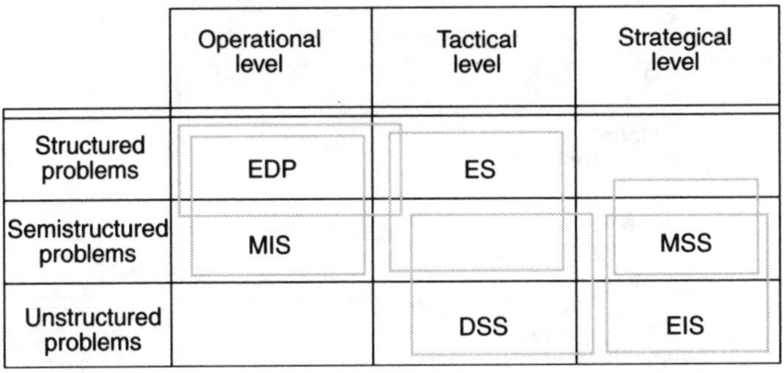

	Operational level	Tactical level	Strategical level
Structured problems	EDP	ES	
Semistructured problems	MIS		MSS
Unstructured problems		DSS	EIS

Figure 7:3 Domains of different computerized decision support systems.

From Figure 7:3 it is clear that DSS is especially adapted to support decisions at the tactical level and in some cases at the strategic level. Observe that an ES is not suitable for solving unstructured problems; this type of problem occurs frequently at the strategic level and requires human judgement and skill at the moment of decision.

Decision support systems are either data-oriented or model-oriented. *Data-oriented* systems could be, for example, on-line budgeting systems while *model-oriented* systems can be exemplified by an accounting system which calculates the consequences of a particular action. When designing systems for structured tasks (MIS) the approach is mainly analytic, planned and deductive. The creation of systems for more loosely structured problems (AI systems) can be characterized as inductive, synthetic and trial-and-error based.

The general structure of a system is shown Figure 7:4.

The user interface for the dialogue management unit consists of a workstation with a set of programs which manage the display screen. It obtains input from and sends output to the user and translates the user's requests into commands for the other two units. The model management unit contains models of the business activity. Examples are spreadsheets, financial models, and process simulation models. This unit also creates, modifies, and invokes the models. The data management unit maintains the internal database and interfaces other sources of data from external databases.

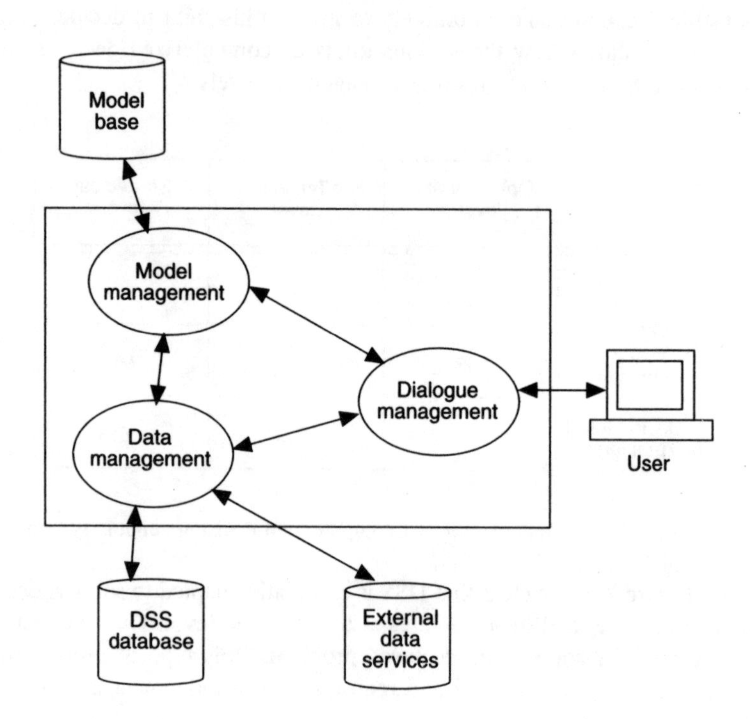

Figure 7:4 General structure of a decision support system.

C^3I systems

This heading is an acronym for **Command, Control, Communication** and **Intelligence**, mainly used as a general concept for a military command structure. The acronym has its origin in the notion of *command & control* (C^2) which has been defined as:

> 'In general terms, C^2 is everything an executive uses in making decisions and seeing that they are carried out; it includes the authority accruing from his or her appointment to a position and involves people, procedures, equipment, and the executive's own mind. A C^2 process is a series of functions which includes gathering information, making decisions, and monitoring results. A C^2 system is a collection of people, procedures, and equipment which support a C^2 process.' (***Coakley*** 1991)

Lately, this concept has been used more and more in a civilian framework synonymous with 'management and decision-making'. Although a certain similarity with normal management decision making always exists, military decisions are often made under quite extreme circumstances including fear for one's own life. The design of military decision support therefore in general emphasizes different qualities from those accentuated in the corresponding civil support. An example is the ever-present demand for higher decision speed. Something has to be discovered, reported, processed, presented, and a decision to take an action has to be made. If this is to take hours or even days, as earlier, the modern battle is lost before it has even begun. Therefore, certain qualities must be present in a C^3I system. These qualities are listed below.

- The system should be specially designed with regard to the needs of the limited human capacity.

- The system must relieve the decision-maker of physical effort and counteract the effects of tiredness. The system should be able to 'remember'.

- The system must strengthen the self-confidence of the lone decision-maker by amplification of his strong points. It should be possible to adapt to the idiosyncrasies of various decision-makers.

- The system should counteract an adaptation of reality to already existing plans.

- The system should interpret reality without distortion.

- The system should amplify signals and trends normally too weak to be noticed.

- The system should enhance systematic interpretation of a massive data and information flow by filtering and sorting.

- The system must enhance instant exchange of information in all known forms without barriers.

- The system must be able to manipulate time, to 'look into the future' by simulation and the making of prognosis.

As a means for understanding the commanding process a standard model has been developed (***J. Lawson*** 1978) according to Figure 7:5.

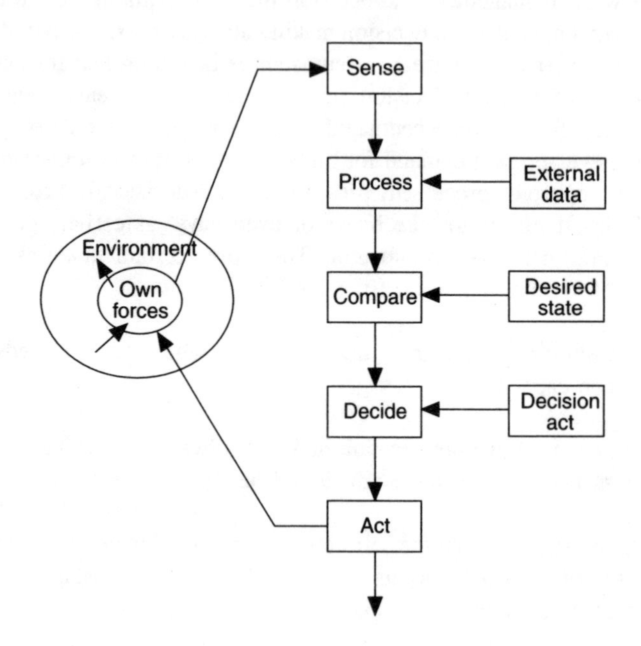

Figure 7:5 Lawson's model for command and control.
(From J. Lawson, 'A Unified Theory of Command and Control', 41st Military Operations Research Symposium, 1978).

The model identifies five functions as follows:

- **Sensor** The sense-function collects data regarding the environment: own and enemy forces, terrain, weather, visibility, etc.

- **Process** The process-function places together and extracts data in order to give the commander information concerning the actual situation.

- **Compare** The compare-function matches the current situation in the surrounding world with the desired – that is, the commander's desired – situation.

- **Decide** The decide-function chooses from the repertoire of possible alternatives for action in order to influence the situation towards that which is sought.

- **Act** The act-function transforms the chosen decision into action.

One of the advantages of the model is that it describes a complex, iterative and dynamic process in a simple manner. It has an obvious similarity with the basic cybernetic control cycle (see p. 55) where the regulating function compares a real value (is-value) with a desired value (should-value). From this comparison the regulated entity is then adjusted.

Another standard model for presenting in this context is the generalized information system model by ***H. Yovits*** and ***R. Ernst*** from 1967. This model includes the influence of the environment, a distinction not made by Lawson's model which places the command/control process outside of the environment of the actual system. The Yovits/Ernst model is presented in Figure 7:6.

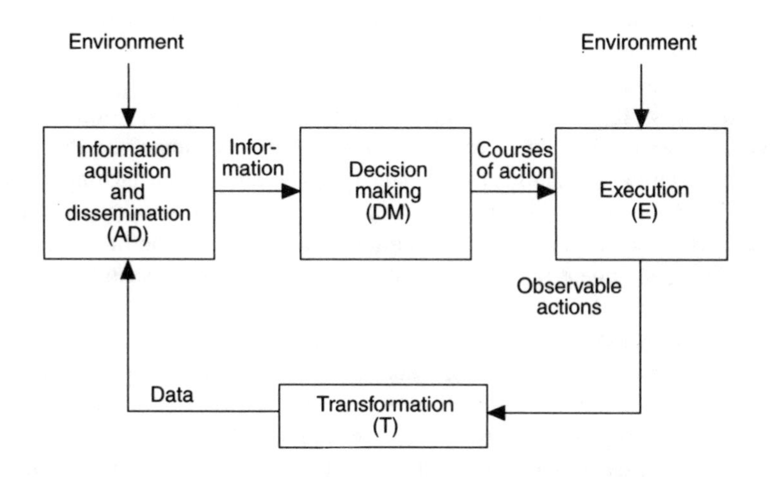

Figure 7:6 The generalized information system model.
(From H. Yovits and R. Ernst, Electronic Handbook of Information, Thomson, Washington DC, 1967).

The demands on the supporting functions provided by a C^3I system can be analyzed with a starting point in the Yovits/Ernst model. **INFORMATION ACQUISITION AND DISSEMINATION** is the function which provides necessary information for decision makers at various levels, building the basis for decisions. To be fit for use, the information has to fulfil certain quality demands:

- **Actuality** The information has to be timely; however, this does not imply that all kinds of information have to be current. A geographical information system managing a fleet of distribution vehicles has to be updated several times per minute, whereas the internal account of the fleet must be updated within twenty-four hours. In turn, the store-value of the enterprise has to be updated once a month.

- **Validity and correctness** It is important that the system's delivery of information to the decision-maker is what is demanded and needed. Too much information can lead to a critical part of it being drowned in the general 'noise'. Too little information implies that decisions are made on uncertain grounds. The information must also be correct. One problem which must be solved is to make the system select and reject incorrect data and estimate the uncertainty of what is presented.

- **Coherence** To have the system work efficiently, common information has to be coherent, that is, it must be interpreted in the same way by all users. This requires a common formation of concepts, where all involved agree upon the significance. Moreover, the form of the information exchange must build upon a standard.

- **Availability** It must be possible to arrive at the information (physical availability) and it must be presented in a way that is comprehensible for the user (cognitive availability). The physical availability can be improved by distributing the system and making the subsystems autonomous (locally needed data should be stored locally). Cognitive availability can be improved if the actual user may determine what kind of data should be presented, and how.

DECISION MAKING seems to an ever-increasing extent to be more complex, and the available time for decision decreases in step with the growing velocity of changes in the environment. The person who can reduce his decision time without renouncing the quality of the decision will be one step ahead of his competitors. With scarce resources, the importance of intelligent distribution and good management will then be more pronounced. This is true no matter if the struggle is fought in the battlefield or in the realm of the business world.

When improving the decision-process, the following must be kept in mind regarding C^3I systems:

- By use of expert systems as part of a C^3I system, knowledge can be accumulated and be available for decision-makers on all levels.

- Simulation improves the possibility to predict the effects of different alternatives of action. New alternatives of action can be tested, which are normally not possible to execute in reality due to the risks involved. A commander can simulate unconventional methods and break established rules without jeopardizing human lives and costly equipment. This possibility stimulates creativity.

- Complex decisions can be divided into manageable parts and be distributed among subordinates. Delegation and decentralization are enhanced.

- The firm establishing of orders is enhanced if drafts can be distributed to decision-makers at all levels and viewpoints can be gathered.

- The risk of under-optimization concerning one's own decisions decreases when overview and holistic judgement become possible.

EXECUTING implies that decisions already taken are transformed into action and that arrangements made are followed up. A well designed C^3I system has the following advantages:

- Decisions can be more rapidly distributed to executing subordinate units. Improved transmission safety enhances the delivery of correct orders.

- In the important phase of execution, the system bridges over individual differences and adapts to the various needs of disparate commanders.

TRANSFORMATION is the necessary feedback process vital for all real-time information systems.

- Instant feedback via the system improves the communication process and the immediate study of the consequences of orders given. As a result better decisions are made.

- The system enhances learning by experience through its logging facilities. The steps in the decision process are documented and the idea behind the decision becomes visible.

Some psychological aspects of decision making

In the preceding pages some structural and technical points of decision making have been discussed. The importance of the psychological factors involved cannot, however, be neglected. One of these is the creative element which more or less influences all steps of the process.

The creative process itself usually proceeds through the following stages.

- **Saturation** the familiarization with the problem and all activities and ideas associated with it.

- **Deliberation** the challenging, rearranging and illustrating of ideas from a variety of perspectives.

- **Incubation** the disengagement of conscious effort, allowing the subconscious mind to work.

- **Illumination** the sudden advent of a bright idea as a potential solution to the problem.

- **Verification** the clarification, reframing, and presentation of this idea in order to obtain other people's viewpoints.

In order to expand the amount of alternatives available, a number of methods enhancing creative group thinking have been developed. The objective of *brainstorming* is to free group members from self-criticism, criticism of others, and inhibition when generating ideas. Group members are permitted to present ideas as rapidly as they occur, without criticism. Freewheeling and wild ideas are welcomed; quantity is wanted and the greater the number of ideas, the greater the probability of a really good one. Combinations and improvement of already existing ideas are also encouraged.

An equivalent to brainstorming is *brainwriting*. It has the same premises as brainstorming, but the exchange of ideas is done on an entirely written basis. By this means, the over-influence of verbally dominant persons in group meetings is neutralized. Brainwriting is especially well suited in a framework of computer-conferencing.

Another popular method is *synectics* (from the Greek 'fitting together'). This technique postulates that creativity exists in every person and that emotional and non-rational factors are as important as the intellectual and rational. It uses an ongoing shift between a rational analysis of the existing problem and the search for non-rational analogies. The more unlikely the analogy the better; improbable analogies will increase the probability that the problem solution has not been thought of before. Often synectics permits persons of wholly different backgrounds to communicate better than by brainstorming. The method is also generally better for dealing with more complex and technical problems.

Good decision-makers who apparently take exactly the right decision at exactly the right time are often asked how they do it. Their answer is remarkably often that they quite simply were lucky. Why some people seem to be more lucky than others has been the interest of the American scientist ***James Austin*** (1978). He presents four general levels of chance in the following hierarchy:

1. Chance happens
2. Chance favours those in motion
3. Chance favours the prepared mind
4. Chance favours the individualized action

The fact that blind, lucky chance happens, is something which can occur for everybody according to plain, statistical randomness. Austin states that this kind of chance is less prevalent than is normally believed. Most lucky chances belong to the higher three levels met by skilled managers or scientists.

Chance of the second level is a kind of luck which is a consequence of sheer curiosity and a will to experiment and investigate. The third level demands a special kind of personal properties and background conditions of knowledge and experience. To observe, remember and create new combinations often invites new chances in a way which is difficult for the particular individual to explain.

The fourth level is dependent on the special individual. A certain combination of interests, life-styles, and lines of thought involve a predisposition to unique insights and innovations – what we call good luck.

Man as a decision-maker has some typical shortcomings. He has a certain tendency to interpret data in favour of himself. Deep-seated models of thought are not readily changed as we do not want to confront new circumstances that were not in accordance with our expectations. Man left alone with himself looks for confirming data, avoiding that kind of information challenging notions that he already has. All this taken together therefore tells us that much is obvious – but only for ourselves.

Perhaps mankind's shortcomings are most visible when it comes to the estimation of probabilities in a decision process. The following subjective treatment of probabilities is well known:

- A tendency to overestimate the occurrence of events with low probability and underestimate those with high probability

- A tendency to believe that an event which has not occurred for a while is more probable to occur in the near future

- A tendency to overestimate the true probability of events which are positive and underestimate those which are negative

An individual forming part of a team of decision-makers sometimes runs the risk of submitting to group pressure. In such cases his own apprehensions are supressed in favour of the collective opinion which often embraces a distorted view of reality. As this is difficult to be aware of, it can be called a *cognitive deficiency* – often the root cause of a disastrous decision. A victim of cognitive defects has repeatedly in history been the messenger who lost his head after presenting the bad news.

Apart from the common foibles in decision making, a number of more articulated pathologies exist. Five such pathologies have been discussed by ***D. Dörner*** (1980) according to the following:

- **Thematic wandering** implying that the decision-maker gives up one goal for another, sometimes several times one after another. Most often the reason is the lack of a general impression.

- **Encapsulation** which is the opposite of thematic wandering. Here the decision-maker concentrates on one goal, being apparently incapable of seeing other alternatives.

- **Refusal to make decisions** originating in decision agony. This can have many psychological reasons.

- **Inconvenient delegation** where the decision-maker lacks the ability to correctly distribute the subtasks to his assistants.

- **Laying the blame on somebody else** when the decision goes wrong.

These pathologies can be divided into two groups. One group includes the first two items and emanates from the inability to define suitable goals. The other group consists of the last three items and is characterized by the inability to learn from experience.

The future of managerial decision support

It is very difficult to provide relevant predictions in such a dynamic field of computer technology as decision support systems. The small examination of the past made here may, however, give some reasonable conclusions for the future.

The users of today's systems differ substantially from the past and are no longer the hard-core computer scientists of yesterday. In the corporate world the use of computers is now a matter of course and has become an integral part of the scene. User friendliness has been made a reality by systems which are neither difficult to learn nor complicated to utilize. Middle-level decision-makers will take advantage of this, personally gathering both internal and external information from the system, in order to make themselves independent of subordinates.

This augmented control of the business environment and the possibility to use new 'last minute information', will give decision-makers better conditions for prompt and substantiated decisions. The prerequisites for direct management and control will be considerably improved.

Another major trend will be the integration of the different kinds of systems. DSS, EIS and ES will all be integrated into each other in one all-purpose EIS system (Everybody's Information System!) which implies that personnel at many levels will use it.

On the other hand it is still difficult to imagine that top level executives should have any major benefit from the system. From the beginning, decision support systems were intended to be a computer-supported tool for the highest level decision. It was, however, soon realized that this category of user would never accept a place in front of a computer screen. Decision-makers at the strategic level have always obtained the information they demanded and they will continue to get it delivered from their subordinates.

To apply the rulebook is something which the computer today tackles successfully and even a robot can learn from experience. But the best decision support will continue to be what is particularly human, namely good intuition, good guesses and a certain feeling rather than slavishly applied rules. No doubt this will also be true in the coming century.

Review questions and problems

1. Describe the real difference between normative and descriptive praxeology.

2. Why is 'good intuition' extra valuable when making decisions under strict uncertainty?

3. What does it imply to dissolve a problem? Has redefinition something to do with dissolution?

4. Present your own examples of structured, semistructured, and unstructured problems in all of the three decision levels.

5. How can a general-purpose computer without specialized software be used as a decision aid?

6. What is the main difference between civil and military decison support systems?

7. Justify why you should perhaps prefer brainwriting instead of brainstorming at your company's creativity meeting.

8 Informatics

- **Electronic networks**
- **Fibre optics, communication satellites, and cellular radio**
- **Internet**
- **Virtual reality**
- **Cyberspace and cyberpunk**

'The success of any system can only be measured in terms of the satisfaction of the user.' (*K. Samuelson* 1977)

Since the late 1960s, computer science has been an important topic of study in all major universities around the world. This subject has its origin in the pure technological areas of computer engineering ('computer science'), computer mathematics ('computing science') and numerical analysis, etc. *Informatics*, with its roots in computer science is a concatenation of the words information and technology. Enjoying features of both, it sits between science and technology and has a broad interface to neighbouring areas such as those shown in Figure 8:1.

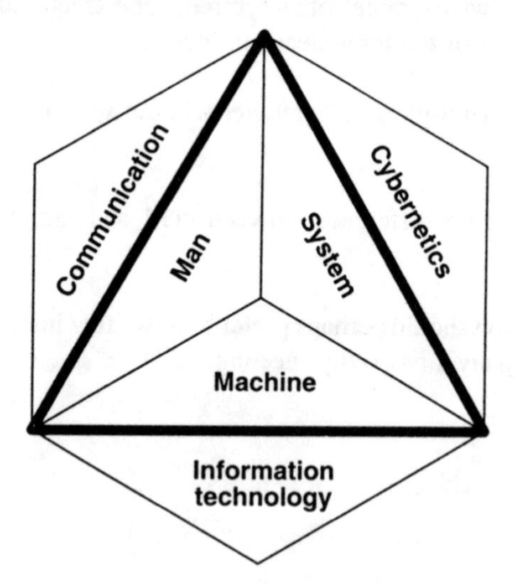

Figure 8:1 Informatics and its intersection with related areas.

As an *area of design*, informatics concerns the overarching construction of information technology, IT, adapted to various user areas. The domain includes supporting methodology and techniques for the analysis, development and management of information systems. As such, it covers the whole spectrum from social science to mathematics. In a sense informatics has more in common with other sciences of design (e.g. architecture) than with its formal origin. Design concerns the shape, the ergonomics and the final user-friendliness of the actual system. Here, technology alone cannot be allowed to decide; the final result must be what customers and users want. For example, the needs of a general user will be widely different from those of a computer scientist working in academia.

Thus political, social, cultural, ecological and especially psychological considerations must be taken into account. A typical problem that involves informatics relates to transportation in medical care. Should the patient be transported to the practitioner, or should the practitioner be transported to the patient? This question can be solved by tele-medicine.

As an applied systems science concerned with user requirements, informatics must necessarily put concrete and pragmatic knowledge before general theories. Therefore, values and preferences are thoroughly dealt with, being the basis for human technological needs. Questions regarding threats to human integrity are a priority concern in informatics.

The most characteristic feature of informatics is, however, how often and crucially the area changes its focus. While social philosophers speculate about the consequences of a certain information technology, the development has already moved in another direction. By tinkering, users discover new areas of use, something not intended by the designers. No doubt, the use of modern IT will transform our present society just as thoroughly as the the industrial revolution changed the society of its time. This transformation towards an IT-based knowledge society has been called 'The third Wave' (Toffler 1980).

In such a revolution, some specific circumstances must be observed. One is the difference between ordinary mail and electronic mail, E-mail (including Fax). With ordinary mail you try to reach a *place*, while by the use of E-mail you try to reach a *person*. Another is a growing need, not for more, but for less information - delivered, however, with an absolute freedom of choice.

In order to make the vast sphere of informatics presentable in a book such as this, the area discussed here is restricted to some major parts of *tele-informatics*, an area of great current interest. Tele-informatics includes knowledge about electronic transference systems for data and information of all kinds. French-speaking countries have coined the name *télématique*, while Scandinavian countries use the name *telematik.*

The international background of today's informatics is an ongoing intense period of development where three global multimilliard dollar industrial sectors are just now merging. These are the telecommunications industry, the electronics industry and the media/entertainment industry. The integration exists on two planes, on the one hand by merger and acquisition and on the other by technological integration of voice, picture, data and information.

One of the main reasons behind the technological integration is digital compression, making possible vastly more 'traffic' on the existing tele- and data-communication networks. Simultaneously, the networks are being rebuilt and augmented into something described as the 'electronic highway'.

Electronic networks

The modern world of today consists to a great extent of an artificial milieu created by human beings and structured into large technological systems. The global telecommunications network belongs to one of the most complex of these systems, even if one only considers the physical network with its telephones, exchanges, optical fibre cables, satellites, etc. If, as well, other parts of the system, such as development laboratories, institutes of technology and general technological knowledge are included, this system takes on gigantic dimensions. Despite its magnitude, it holds together and forms a working entity with an impressive level of reliability in service. This reliability in turn, may to a great extent be ascribed to 'self-healing qualities' with on-line, automatic fault isolation and call rerouting.

One of the most interesting properties of electronic networks (apart from the fact that they work so well!) is that they are constantly undergoing change. They are modified by different phenomena, such as technical development, investment, changed legislation, merging of carriers, implementation of new services and connection of new users. The system thus develops according to its own logic, a logic which is itself subject to change and thus changes the whole system yet again. Like many other big technological systems, the electronic network is therefore characterized by an interplay or combination of change and stability.

Another interesting feature of the modern electronic network is that it neutralizes the age-old debate regarding centralization versus decentralization. All the capacity of extreme centralization and extreme decentralization can exist simultaneously in the same network. With regard to this quality, the modern electronic network parallels the human brain, which must be considered both centralized and decentralized at the same time.

Networks consist of *links* connecting the numerous *nodes*. Links are of various kinds, from simple copper wires to transmitting and receiving satellites. The nodes are exchanges, today mainly computerized and often both processing and storing the flow between the links. Distinctions can be made between the following networks:

1. Information networks
2. Communication networks
3. Computer networks
4. Relational networks
5. Hybrid networks (of 1, 2, 3, 4)

Information networks are characterized by nodes which are databases, while communication networks connect information processors (human beings or computers). The communication system enjoys a development which is relatively independent of the information system that it supports.

Two basic styles of network exist: the *hierarchical* in which one node or exchange is designated to manage the other, and the *peer-to-peer*, where every node manages itself. In the second case, all devices are of an equal status and there is no hierarchy for communication. With the combined intelligence of all nodes in the network, a peer-to-peer configuration is able to cooperate and automatically to determine the best routes and also reroute around problems. This capability is the result of a continuous sensing, feedback, feedforward and adjustment among all subsystems communicating within the network.

Each kind of network has its own morphology and evolutionary pattern with a different appearance of nodes and links. Complexity, reliability, vulnerability and dependability vary according to the type of network. Evolutionary growth, rather than planned development, is a characteristic property of nearly all networks and has been studied by several communication researchers, among others Samuelson (1977). The growth is subject to the needs of its users and is largely independent of existing technology.

The various stages of this growth can be considered a life-cycle with the steps shown in Figure 8:2. The stages are:

1st stage Differentiated loci exist randomly distributed within all dimensions of a medium

2nd stage Formation of interconnections by 'Brownian' motion

3rd stage Choices are being made; the dynamic is no longer Brownian

4th stage Limiting the alternatives

5th stage Limiting the number of connections and the emergence of decision bias

6th stage Increasing structure and differentiation of structure

7th stage System fragmentation prior to renewal

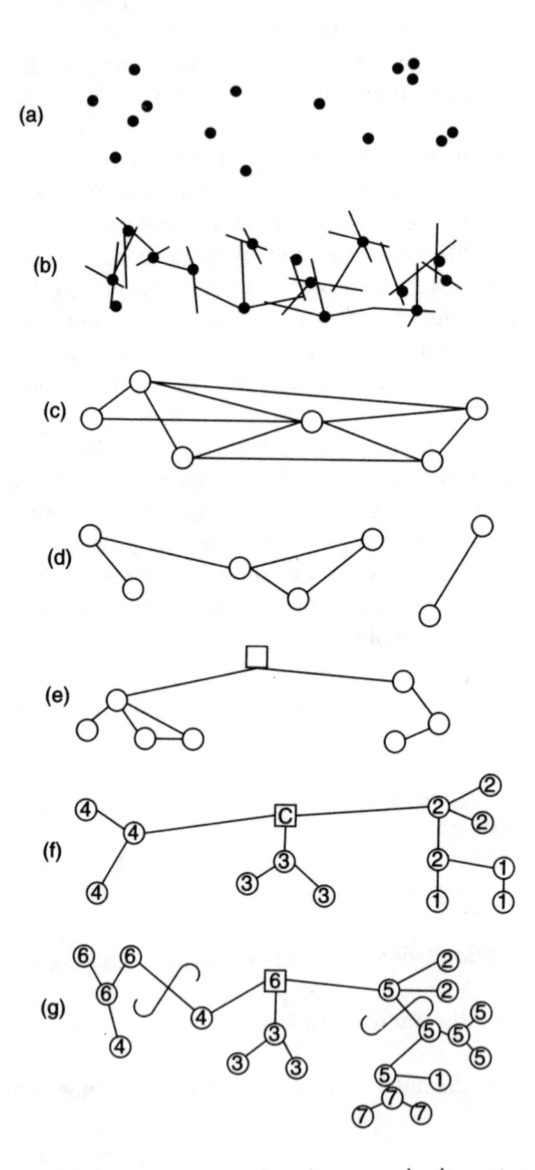

Figure 8:2 Life-cycle of a telecommunication network.
(From K. Samuelson *et al.*, *Information Systems and Networks*, North-Holland, 1977. Reprinted with permission.)

Figure 8:3 shows evolutionary growth and orderly expansion in an information network as compared with that in a communication network. Note that there are only six stages, as the initial networks already exists.

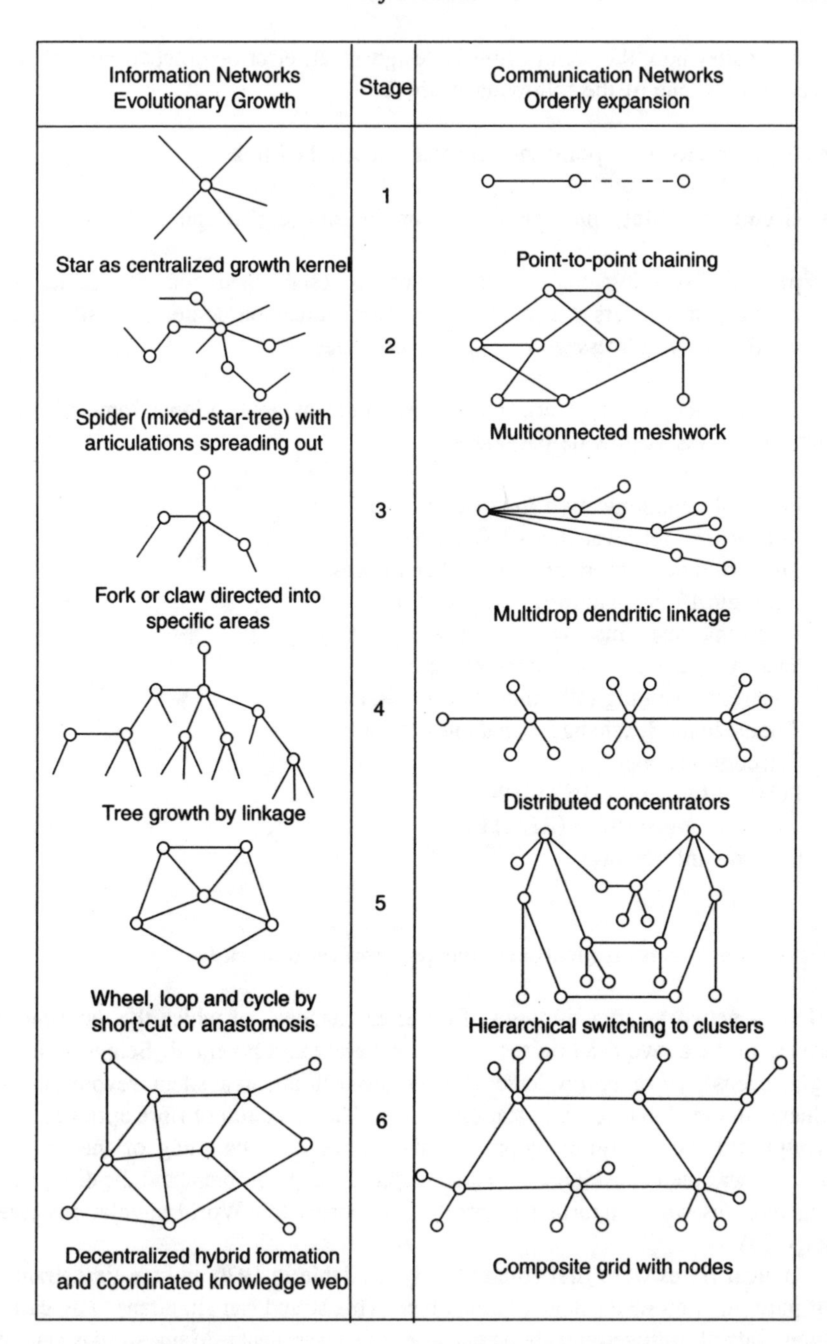

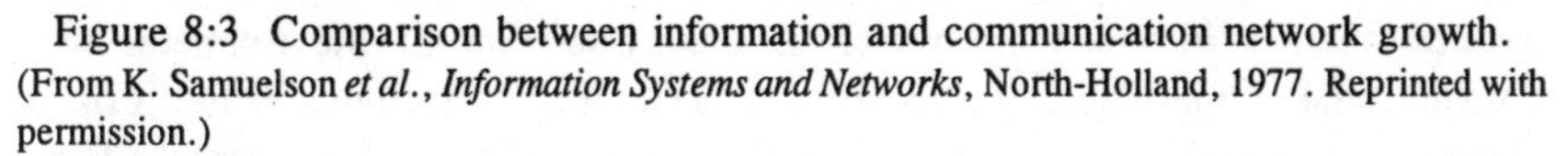
Figure 8:3 Comparison between information and communication network growth. (From K. Samuelson *et al.*, *Information Systems and Networks*, North-Holland, 1977. Reprinted with permission.)

No matter how the architecture is designed, all electronic networks function according to one of the following methods:

- **fixed connection** point to point via a subscribed link

- **circuit switching** physical linking on demand (e.g. telephone line)

- **packet switching** logical connection is established via the nodes of commercial carriers which relays the communication content, consisting of small electronic message modules or packages.

The advantages of networks may be seen in many applications. Some of those that exist on a worldwide basis are:

- Personal communication (telephone, fax)
- Information retrieval (knowledge databases)
- Meteorological information (weather forecasts)
- Geological information (earthquakes, etc.)
- Electronic mail (message storage and forwarding)
- Sale of tickets (air-carrier tickets, etc.)
- Electronic banking (all kinds of commissions)
- Telemedicine (exchange of patient data, etc.)
- Teleconferencing
- Police information (INTERPOL)
- Financial information (REUTERS)
- Customs information

Fibre optics, communication satellites, and cellular radio

Modern networks are composed of different channels, of which the three mentioned in the above heading seem to have the greatest potential. Being the least 'glamorous', *fibre optics* have already brought about a silent revolutionary change in the global communication system. The invention of fibre optics in fact neutralized the emerging 'copper crisis' during the beginning of the 1970s. Copper was then considered a serious global shortage problem and the diminishing reserves were important parameters in Forrester's World simulations (see page 26).

Optical fibres were first commercially available in 1970 and are tiny strands of pure glass no wider than a human hair. This strand can simultaneously carry thousands of digitized telephone calls or their equivalents. Many strands can be joined together into a cable which is only one-sixth of the diameter of a conventional copper cable. Information travels along the strands in the form of extremely fast light pulses zigzagging inside the cladding of the strand. These pulses travel at a higher frequency than electric current and are generated by laser diodes, pulsing at a rate of many million per second. Technically, the process is called PCM or pulse code modulation, which implies that a quantified analogue signal is transmitted in digitally coded form by means of light pulses (see the sampling theorem on page 153). At the receiving end, the light is converted back to an analogue signal.

Compared with ordinary copper cables, optical fibres have a number of advantages, including the following:

- Higher capacity and speed
- Cheap to produce (raw material is sand)
- Light and easy to handle (takes little space in crowded underground ducts)
- Need few repeaters to amplify the light pulses
- Free of electrical interference
- Practically impossible to tap

From a relatively slow start the progress of installing fibre cables is now accelerating. Several transatlantic submarine optic cables requiring repeater stations only every 150 km have already been laid. These cables compete with satellite links and it is expected that their cost-efficiency will soon surpass that of a corresponding satellite system covering the same distance.

Communication satellites are another technology that has made possible the telecommunications revolution. Such satellites have a geosynchronous orbit 36000 km above the equator which makes them appear to be stationary when viewed from earth.

Something which the world seems both to need and like is terrestrial *cellular radio telephony*. The widespread use of mobile phones has already revolutionized the global communication pattern. Cellular radio works by dividing city areas and counties into transmission zones called *cells*. In every cell there is a low-powered transmitter with a range adapted exactly to the cell. This means that transmitters in cells not adjacent to each other can share the same frequency; a kind of frequency reuse.

A global system called TELEDESIC with 840 satellites in low orbit has been proposed by the famous computer entrepreneur Bill Gates and his colleague Craig McCaw. This system should be a substitute or successor of conventional cellular telephony supplying the world with (besides telephony): fax, data communication, staff locating, and position fixing. The whole project would involve expenditures of about ten billion dollars and give a connection to literally every place on earth. From a technical point of view there is now doubt that the project can be successfully realized. However, it must be questioned whether the world needs such a system, even if there were enough subscribers and it could be financed.

Internet

When examining electronic networks, the Internet is a good example, as it is one of the biggest and most spectacular. It is possible to compare the development of the Internet with earlier, radical infrastructural changes caused for example by the arrival of the railway or the motor car. Viewd in this way, the Internet is a logical continuation of a development that began with the wire telegraph and continued with the telephone. As a hybrid network it exhibits all the phenomena associated with both orderly expansion and evolutionary growth.

The Internet grew out of a project initiated by the American Advanced Research Project Agency - ARPA, in the 1960s. This agency was subordinate to the US Department of Defense and its mission was to create a big network for communication between military research computers. The network acquired the name ARPANET and soon came to include many universities pursuing military research. The location of its main node became Washington DC.

The original concept was to make the network resistant to a nuclear war. One or several nodes could be destroyed without devastating consequences for the network as a whole, thanks to its distributed character. The big innovation was that data transmission was based on 'packet switching'. This technique divides the information content of a message into small electronic packages of equal length, each equipped with an address tag. Each package could be routed on different ways in the network, a very practical feature in the case of bottle-necks or breakdowns.

As more and more universities from all parts of the world were connected to the network, its character changed to that of an international university communications facility. The network soon become a conglomerate consisting of the different nations' own networks and was taken over by the American National Science Foundation under the name of the *Internet*. It can now be characterized as a 'network of networks'. For experts in the area the term for all integrated networks of the world is the *World Net* or *The Web*. The Internet, however, must be regarded as the core of this phenomenon with its more than thirty thousand individual networks.

The Internet is today an example of natural, 'anarchistic' growth from below, where the wish to connect was the only main shaping factor. There is no board of directors or central command authority for the network; from several points of view it manages itself. It can thus be characterized as unhierarchic, uncentralized, and unplanned. After the modest start at the end of the 1960s, the Internet has exploded and is now the backbone of global data communication with its more than 30 million computer users. Furthermore, this group is growing every year and today embraces 92 countries. From a technical point of view, the Internet is a digital highway with a capacity of 34 Mbits per second on its main links. It is a typical peer-to-peer network where the local branches have a capacity of 2 Mbits or below. In a global network of such dimensions, certain problems emerge of necessity. These include questions about the kind of information that may be distributed, and the 'human information rights' of the participants. For example, a special group has been founded to protect human rights in the electronic world called the Electronics Frontier Foundation (EFF). Politicians and lawyers are continually trying to figure out whether to govern or police the network and if so how. Authorities speak about legally responsible publishers for bulletin boards, censorship to prevent the distribution of pornography, and the prohibition of information encryption.

Cryptography on the Internet is a problem area of its own. Much of the information transmitted on the network must be considered attractive for various reasons. Business information, stolen records, military secrets, passwords and private mail are constantly available for those capable of reading it. Probably there are no safe systems on the Internet and computer security experts state that at least one million passwords are stolen every year.

The need for electronic privacy therefore has prompted a growing use of cryptography. In United States the official attitude to this trend is said to be positive – on condition that the authorities hold a key to a 'trapdoor' in every system. This is the idea behind the criticized Clipper system which holds the official standard for information encryption. Other countries, such as France, have a still stricter attitude towards cryptography; there, all cryptography should be licensed, a legitimate reason for its use established and copy of the key delivered to the authorities.

The fact that many new digital technologies like cellular telephones are untappable lay behind the new US Digital Telephony Act. This law states that telephone companies must use communication software which permits the authorities to tap and read the bit-stream. A *cause célèbre* is, however, that the Clipper has been partially cracked by a scientist from Bell Laboratories in New Jersey.

The business world, together with the majority of other users of the Internet, are rather cool *vis-à-vis* all attempts to share codes and keys with 'Big Brother'. The idea in itself makes secrecy a corrupted concept and many have turned to what is called public key codes or PKC. In contrast to classical codes this has two keys. One of them, the public key, is given to the person who wants to send information to you and is used for encoding. The other is an unrelated 'private' code key used by the receiver to decode the message. It resides in the personal computer where it is less likely to be stolen. If someone get hold of the public code nothing can be done with it. Neither the message nor the public code itself tells anything about the private part. (Consequently, some people distribute their public code key on the network.)

The use of codes in itself indicates that certain information is kept secret for other persons. There are now on the Internet more discreet methods which even hide the fact that something is hidden. They may be characterized as methods of 'hidden meaning'. Hiding messages in innocent text and pictures has long been a technique used; doing it in the relevant digital bitstream is both easy, cheap and undetectable. Within the millions of bytes representing everything from sound files, high resolution pictures, private letters or financial transactions, every kind of secret information can be hidden. Even cryptographic experts admit that such hidden messages rarely leave enough of a pattern to be detected and decoded.

Serviceable and widely available encryption algorithms are the prerequisite for something which can be called *digital signatures*. This is a piece of code which has its origin from a certain person or rather his personal computer. If such a signature can be considered entirely trustworthy it can be used to confirm and acknowledge all kinds of transactions in a network. Such hidden transactions could easily be the backbone of an alternative economy unrestricted by governments and a nightmare for assessment authorities.

It is, however, doubtful if it ever will be possible to regulate the Internet, because what is forbidden in the US will be allowed in Finland, and so on. Also for purely technical reasons, it is at present impossible to censor the network, since it is constructed to work around blockages and censorship by its self-repairing qualities.

Of course, there is a real danger that the Internet may be threatened by the very qualities that supported its growth. It is influenced from all sides; veteran users try to protect it, governments want to control it and pornographers try to exploit the freedom of it. The greatest current threat seems to be the commercial interests which strive to make money on it. If they were to grow too strong they could cause a sudden collapse of the Internet – one of the most promising cultural phenomena of the 1990s.

The earliest applications on the network have been electronic mail (E-mail) and computer conferencing. Classified as electronic mail is everything which is not ordinary mail ('snail mail') such as telefax, videotext, searching in external databases, conferencing, etc. The use of E-mail is in fact much more rapid than the use of an ordinary fax device. In addition an E-mail is safer from wire-tapping during transmission and quite private in contrast to the fax message which can be read in public at the receiving end. The need for E-mail is one of the major reasons for the growth of the Internet.

Although E-mail is still one of the most used applications, practically everything which can be transformed to an electronic bitstream can be communicated via the Internet. Political debate, literary criticism, stock market tips and matchmaking are some examples. Archives, libraries and databases are available around the clock, and very often without fee. Today, a researcher can publish a report on the Internet and receive an immediate response instead of waiting several months.

Many academic institutions around the world have their papers and publications stored in databases retrievable for both researchers and the general public. The latest highlight is that *Encyclopedia Britannica* has found its way on to the Internet. It will be accessible through MOSAIC (see p. 239) and tests have been carried out at the University of California. Some 44 million words and many thousand of pictures are stored in a database. Free electronic magazine subscriptions are common on the Internet. One example is the Internet gazette *Refractions* which contains news about various electronic forums. There are already more than 3000 forums for different discussion topics.

Some of the main areas of information accessible on the Internet include:

- Scientific data, for example star catalogues, earthquake danger zones, research papers, etc.

- Sound and pictures. Internet users digitize pieces of music or pictures and share with others. Today, short digitized moving video sequences are to be found on several databases.

- Electronic newspapers, magazines, books and various kind of fact-files. Project Gutenberg in the USA and project Runeberg in Sweden have digitized several hundreds books which are available free of charge as the copyright has expired.

- Software. Normally this is what is called shareware or other kinds of programs which can be distributed free.

A main criticism of the Internet has been that network services are complicated to use especially when transferring files between universities and enterprises. FTP, or file transfer protocol, has been used to do this for many years. Its user interface is rather primitive, and bears a certain resemblance to earlier DOS versions. To improve matters for the operator and to facilitate the use of the network, a simple graphical interface called GOPHER has been introduced. Instead of cryptic file and catalogue names, longer explanations of documents and archives are given in a graphical environment. With GOPHER and its hypermenu system the user has no need to know on what computer his information is stored. By choosing a certain menu it is possible to navigate throughout the network. Some of the menus connect directly to the requested information which is presented as text on the screen. A university may have a GOPHER menu connected to the computers of many other institutions. There are already more than 5000 GOPHER servers in use around the world.

The newest way of presenting information on the Internet is called World Wide Web or WWW. This is a further development of GOPHER with graphics, pictures, and hypertext. Hypertext implies that specially indicated words or phrases in a text are equipped with links to new information quantities. These links crisscross the network in all directions. WWW has an unique application for windows called MOSAIC which is a navigation system used by pointing and clicking a mouse. It has full capability in handling both sound and pictures. Appropriate software for Unix and Macintosh computers is called CELLO.

In 1995 there were about 1000 WWW servers, most of them at academic institutions. Several private enterprises had already adopted WWW; for example,

Novell were storing their drivers and reference articles on such a server. Another interesting new field of application involving WWW severs is the connection of indicators. Some research institutes in US have coupled their Geiger-counters to the network, thus making continuing registration of radiation available everywhere.

All this makes the Internet more of an intellectual working tool than a communication medium. The refinements of the network have grown more and more and have become increasingly user friendly. Anyone can now sit at his desk and help himself to the accumulated knowledge of the world. The Gutenberg art of printing created the foundation of the modern nation state. The printed word was the glue which joined together the different parts of a nation. In the same way the international networks are on their way to creating a worldwide community.

Although the base of users is constantly broadening, the majority of them are university graduates and persons in private employment who have access to the Internet via their job. It is still rather expensive for a private person to connect via a modem and a personal computer at home.

Many communication enthusiasts now speak of the Internet and its users as the modern 'network society' existing in 'cyberspace' and populated by 'cyber-citizens' following the 'netiquette'. A special kind of 'net-culture' has emerged where researchers, technicians and students exchange ideas and information, often in the form of computerized conferences.

The net-culture is based upon a sort of anarchic ethic, embraced by the students and 'hackers' who took part of the early build-up of the network. Among its implicit basic rules are the following:

- All information should be free
- All access to computers should be unlimited and total
- Promote decentralization and mistrust authority

Much of the information residing in the Internet is in fact free. Accounted as free information must also be all those computer programs which are possible to download for personal use. Many of them enhance access to various parts of the network and may be regarded as its self-organizing agents. For example, by 1995 some ten million copies of MOSAIC had been distributed in 75 countries.

Based on the written word and the English alphabet extended with various punctuation marks and computer characters, the writing style has become a special part of the net-culture. Certain acronyms and signs are commonly used to express frequent expressions and emotions. Some examples are given in the list below (If the head is turned sideways the signs expressing emotions are quite striking!)

Acronyms		**Emotions**	
FAQ	Frequently asked question	:-)	I am smiling
IMO	In my opinion	;-)	I am winking
IMHO	In my humble opinion	:-(	I am frowning
LOL	Laughing out loud	{:-)	I am wearing a toupee
MOTOS	Member of the opposite sex	= :-|	I am a cyberpunk
MOTS	Member of the same sex	:-D	I am laughing
RTFL	Rolling on the floor laughing	:-X	My lips are sealed
RTFM	Read the ... manual!	@:-)	I am wearing a turban

Warning! Do not type in capital letters (IT IS LIKE SHOUTING!).

The small constellations in the right-hand column are called 'smileys' or 'emoticons'. They are used to express the body language and feelings existing in an oral conversation but which are lost in the E-mail system.

Being a two-way medium, the Internet must be considered a many-to-many facility. As such, it has dramatically changed the social rules as to who may talk to whom and who may listen. In fact it is today easy to contact politicians and those in power. Most networkers are aware that they have direct access to the computer screen of America's president, at least in theory if not in practice. The White House now receives about 4000 E-mail messages a week, stored, handled and filtered by a special group of aides. How many of these messages really reach the eyes of the President is an open question. Probably not more than the fingers on one hand as a certain filter technique has been developed to avoid total information input overload. Filters for ordinary networkers who believe that their lives will be both saner and better if they can avoid reading messages from certain detestable individuals - called bozo filters - are common.

Many enthusiasts (including US Vice President Al Gore) have stated that the new network technology will change society completely. The changes are going to be as radical as when the industrial revolution superseded the peasant culture. We may perhaps be about to enter a new democracy like that of the old Athens. But changes of this kind are often revolutionary, beyond parliament. From that point of view, a vice president cannot influence the development very much.

However, even as we speculate about a future network society, the old division between information consumers and producers is already loosening up. Today, anybody can be a producer or a publicist and have at his disposal a mass medium in the network. Historically, distribution has always been the problem, but with modern networks this is no longer the case.

Twenty years ago nobody had a feeling that the computer, which was then the controlling instrument of 'Big Brother', would be a tool for freedom and democracy. Thanks to the networks we now have the whole world within reach of our index finger. Twenty years ago, the only persons with the world within reach of an index finger were the American and Soviet presidents and then in a very devastating sense.

Virtual reality

In the same way that a telephone is a tool to facilitate conversation over long distances, the technology of *virtual reality,* or VR, is primarily a facilitator of a total remote communication experience. A higher, more effective level of communication is now possible by the amalgamation of imaging and computing power. This is something which has opened up the possibility for shared expertise and experience in a way which was impossible to imagine only a few years ago.

Behind the concept of VR lies the fact that human brains work considerably better through sight, sound, touch, and smell instead of just with text and numbers. Of course, there are different levels of experience involved when just looking at a picture of an aquarium, looking at a real aquarium or putting on scuba gear and swimming in it.

Besides better shared real events, VR has made possible virtual experiences that are impossible or too dangerous in real life. It may even extend the boundaries of our senses beyond what we have experienced earlier when the user takes an active, participatory role in a world created by the computer. The French author *Marcel Proust* once said that 'the real voyage of discovery consists not in seeking new landscapes but in having new eyes'. In our time those new eyes are located inside the VR helmet experiencing this phenomenon called *virtualization*.

Among the qualities of the VR is the absence of all physical limitations belonging to the real world. Gravity, speed limits and other phenomena dependent on the laws of nature can be disregarded in the virtual world. Virtual things can be examined from every possible angle and be enlarged or reduced to a degree only dependent on the speed and power of the computer being used and the quality of the model. Basically, VR is nothing more than a three-dimensional computer simulation where the executor is situated inside the image. From this position he can change the direction of view and observe new parts of the simulated surroundings. He can also twist, turn, and move objects which are part of the image.

To create a Virtual World it is necessary to have at one's disposal a very powerful computer and a large database attached to a high capacity communication system. The computer records the movements of the user, calculates the consequences and presents new scenes of the VR which is displayed in the helmet. The computer must have sufficient memory to accommodate a whole virtual world and be so powerful that it can represent this world in two stereographic scenes. Furthermore, the calculations have to be updated so frequently that the user has no impressions of a disjointed world.

An other key component is the helmet which has two purposes. It registers in what direction the wearer looks and it presents the VR. To do this it has sensors which report the movements to the computer and two small monitor screens, one for each eye. Together the screens present a three-dimensional scene of the virtual surroundings. The combined effects of the presented scenes, a built-in stereophonic audio system and the sensitivity for head movements, create the necessary conditions for the user to merge with the virtual world. The fact that the helmet screens off reality facilitates the process. The third key-component is a three-dimensional pointing device which is used to highlight details and to influence the VR. This is also a steering device with which to transport oneself.

VR has a rapidly expanding area of applications within various fields, some of which will be mentioned here. *Telepresence*, which can be described as the projection of a human mind to a remote site, has many industrial and military applications. Telepresence is often combined with the remote control of robotic devices, then called *telemanipulation*.

Imagine a robot working inside a damaged nuclear plant which is emitting strong radiation. It is manipulated by an operator from a place hundreds of miles away. The robot is equipped with TV cameras, microphones and various sensors for radiation, temperature, moisture, etc., and is continuously transmitting information back to the operator. The operator is wearing a helmet provided with a complete audio/video system and tracking devices. Equipped with this helmet and using data-gloves and a steering device, together with adequate feed-back arrangements, it is possible for the operator to feel as if he is in the place of the robot. By turning his head, the operator simultaneously turns the robot's 'head' and gets a new perspective of the interior inside his helmet. The data-glove in turn commands the robot's arm and receives a sensory virtualization of the manipulated object. The operator is now interacting with the remote environment, the system being the medium.

Today commercial aircraft and car manufacturers routinely use VR environments for designing and developing their new models. The new technique is an extension of the old computer-aided design or CAD where three-dimensional drawings are created on a computer screen. Without use of a pencil, the drawings are then manipulated, updated and stored in a database. The whole concept can then be analyzed and even tested in a simulated destruction without actually having been built, something called computer-aided engineering or CAE. By this means, the need for developing scale-models has been dramatically reduced, as has the need for building and testing working systems as prototypes.

Architectural design appears to be especially well suited to merge with VR technology. Here a technique called 'walk-through' has been developed which facilitates cooperation between clients, designers and subcontractors. By access to the virtual construction the coming design can be inspected from inside and any proposals for alterations can be put forward immediately. Of course, all this takes place before any actual construction has begun.

Walk-throughs of quite another kind can be used in VR applications for network maintenance. In a virtual electronic landscape that visualizes the nodes and links of a communication network, maintenance personnel can move around and investigate the switches. Problems can be discovered, data-flows optimized and new connections established before congestion arises. Visualizations that illustrate different properties and flows of the network can be expressed via the width of the links, their colour and excitation. See Figure 8:4.

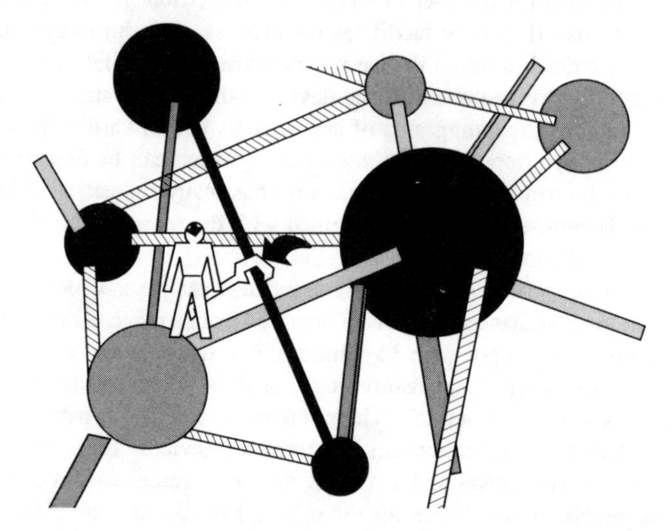

Figure 8:4 Virtual maintenance of a communication network.

Applications of VR in the medical area are already numerous. Telemanipulated microsurgery has been particularly successful where a specialist performs an operation by being virtually inside the patient's body.

Finally, the world of entertainment has to be mentioned. Many video-entertainment centres already exist which have different kind of VR simulators. Here it is possible to experience the working environment of a jet-fighter pilot, a submarine commander or a racing driver. Also arcade games are offered where the participants look into periscopelike devices which give three-dimensional images. In many of these games it is possible to create personalities in animated characters according to one's own wishes. These virtual egos are then put into action as part of various adventures.

Cyberspace and cyberpunk

In 1984 a novel was published by the American author **William Gibson**. Its title was *Neuromancer* and Gibson was soon considered a promising representative of 'New Journalism'. This trend was a carefree, conscious mixture of facts from the latest technological and biological achievements and dystrophic prophesies regarding social development. Here the world was recognized as a worn-out, rotten, dirtied and overpopulated area, controlled by multinational enterprises and housing dubious underground techno-cultures. The text consisted of several newfangled ideas such as:

'The sky above the port was the Colour of television, tuned to a dead channel.'

or

'It was hot, uncommonly hot, beastly hot...the air was mortally still and and the high cloudy sky had a leaden, glowering look, as if it wanted to rain but had forgotten the trick of it.'

Gibson's world is populated by ordinary people, more or less resembling ourselves. However, they can choose their appearance by plastic surgery and bodily protheses. Everything can be bought for money, even length of life. A new dimension was born when Gibson introduced *cyberspace*, a mental room formed by all the interconnected computer systems of the earth. It is a non-place where information is transformed to an artificial computer landscape. Here the users can connect themselves and move freely in a virtual space covering the whole planet. By the introduction of cyberspace a third, alternative world, created from a synthesis of the real world and the dream world was born.

Cyberspace soon replaced outer space as a main area of interest. Its heroes become the future generation of data hackers (called 'cowboys' by Gibson). These connect to the virtual world through an interface linked directly to the brain via a small contact behind the ear. The software loaded in the brain was described as *wetware* by Gibson.

The cyberspace is described as an icy abstract emptiness, integrated with a red glowing check pattern called the *matrix*. In the matrix, the databases of the world are visualized like huge building blocks. The visitors of the matrix travel through this space, looking for and surveying different kind of visualized data. For them the virtual world is more real than the existing physical room surrounding them and endlessly more attractive.

Gibson seems to have hooked on to the technical development at exactly the right moment. Many computer applications seen as dreams only a few years ago, like VR, have now been realized, at any rate in the laboratories. The information flow was already organizing itself into that more coherent structure we now see today (multiple-lane information freeways, international databases, etc.). Cyberspace is unimaginable without the Internet.

During the following years something curious happened. Like a cultural virus, a trend called *cyberpunk* slipped out of the pages of Gibson's novels (and others). A mixture of hi-tech, subcultures, street-attitudes, and digital communication left the realm of the science-fiction ghetto and moved out into real life. From the beginning the term had an appropriate sound. Many cultural wizards began to call themselves cyberpunkers – from computer hackers to designers and techno-musicians. Some scientists missed the very point concerning Gibson's criticism of the new technology and the social transformations caused by the computer. They considered Gibson as a prophet and product developer.

The attraction of cyberpunk seems to be the invalidation of the old Cartesian dualism. In cyberspace the limits between reality and dream have been dissolved. Here man and machine have merged into one unit, into *both* mind and body. The virtual world is regarded as the liberation of human senses. It is a return to the close communication of the old community. Formerly, people spoke to each other from window to window; today they connect to the computer network, the final result of which is the 'virtual city'.

Review questions and problems

1 No general scientific theories can be associated with the area of Informatics. Give an explanation why.

2 How is it possible to speak of self-healing qualities in an electronic network?

3 Try to explain the similarities between the establishment of new social contacts after a move to a new town and the development of a telecommunications network.

4 Should it be possible to park a geostationary communication satellite in a polar-orbiting plane?

5 What was the main design idea behind the origin of the Internet?

6 Give some examples of main problems existing in the Internet from
a) the authorities' point of view
b) the private individual's point of view

7 How can the Internet be said to promote personal freedom in the world?

8 Why has cyberspace replaced the outer space as an area of interest today?

9 Some of the Systems Methodologies

- **Large-scale, soft, and intertwined problems**
- **Systems design**
- **Breakthrough thinking**
- **Systems analysis**
- **Systems engineering**
- **GLS simulation**
- **Method versus problem**

'To find the problem is the same thing as finding the solution.'
(*Rittel and Webber* 1974)

Systems science has assumed as its primary responsibility the task of handling real-world, large-scale, intertwined problems of complex systems. It is applied with the assumption that these problems are of a certain similarity, regardless of the system they have originated from. The systemic view therefore includes a special attention to the threat of technological 'fixes' and pervasive side-effects of the far more common mega-technology which is invading society. Also, the emergent properties, possessed by the system but not by its parts, are extensively involved.

In this category of problems – relating to the social, behavioural and organizational fields – the traditional scientific methods have proved to have substantial shortcomings. Moreover, these methods also seem to pose new complications for the problems to which they are applied. These methods thus have to be replaced or be complemented by new ones, capable of handling the soft variables such as values, motivation and sentiments that are an integral part of all social systems.

As approaches to large-scale problem solving, the methods presented here may be considered as a family of coherent methods when dealing with systems problems. All of them emphasize the interactions and interrelations between the diverse parts of problems. This is in contrast to the fragmented approaches that are so often taken when eliminating symptoms of social and organizational ills. Systems methodologies are generally *systematic*, in the sense that they are composed of rational and well-ordered steps, taking into account the range of probable alternatives or perspectives. The systematic approach ensures that solutions can be *planned*, *designed*, *evaluated*, and *implemented*. In this context the operational definition of 'methodology' has to be compared with the terms 'technique' and 'philosophy' according to the following commonly used definitions.

– Philosophy: a non-specific and broad guideline for action.

– Technique: a specific programme of action, producing a standard result.

– Methodology: more precise than a philosophy but lacking the precision of a technique.

A methodology can be used for two purposes: either to bring a system into being (systems design) or to refine an already existing system. If the latter is done without distinguishing between beneficial as opposed to harmful transformations for human beings, it is called *systems improvement* by van Gigch (1978). He correctly points out that the operation of a crime syndicate can naturally be improved. Systems improvement does not question the function, purpose, structure, or process of interfacing problems. It is often done for the wrong reasons and the solution can be worse than the problems that it was intended to cure.

Large-scale, soft, and intertwined problems

Planners and problem solvers dealing with large-scale societal problems have long been aware that their situations are quite different from those of ordinary scientists and engineers. Classical methods of science and engineering have little if any relevance in their work. They are no longer surprised to find that their solutions often induce problems of greater severity than those that they were intended to solve. By nature, social systems are interconnected networks where output from one part becomes input to another, thereby obscuring the determination of where the real problem is situated and how to intervene. It is also said that social systems have no goals to be achieved, rather they have relations to be maintained.

A main determinant for the complexity of large-scale systems is the ever-increasing rate of change in modern society. This in itself adds to the complexity of all problems, which in turn implies more time for their solution. Russell Ackoff (1981) says the following about rate of change: 'The more the rate of change increases, the more the problems that face us change and the shorter is the life of the solutions we find to them. Therefore, by the time we find solutions to many of the problems that face us, usually the most important ones, the problems have so changed that our solutions to them are no longer relevant or effective; they are stillborn.'

An excellent examination regarding the nature of social problems has been performed by ***H. Rittel*** and ***M. Webber*** (1974). Their main thesis is that social problems (which they call 'wicked problems') are never solved. At best they are only resolved – over and over again. The following main statements are adopted from their work.

- **There is no definitive formulation of a wicked problem**
  The formulation of a wicked problem is the problem! Finding the problem is the same thing as finding the solution.

- **Wicked problems have no stopping rule**
  The problem solver terminates the work for reasons external to the problem. He runs out of time, money, or patience.

- **Solutions to wicked problems are not true or false, but good or bad**
  Many parties have equal right to judge the solutions.

- **There is no immediate nor ultimate test of a solution to a wicked problem**
  Consequences of the solution may yield utterly undesirable repercussions which outweigh the intended advantages. The full consequences cannot be appraised until the waves of repercussions have completely disappeared.

- **Every solution to a wicked problem is a one-shot operation with no opportunity to learn by trial and error**
  Every implemented solution is consequential. Large public works are effectively irreversible; you cannot build a freeway to see how it works and then easily correct it afterwards. The effects of an experimental curriculum will follow the pupils into their adult lives.

- **Wicked problems do not have a defined set of potential solutions, nor is there a well-described set of permissible operations to use**
  There are no criteria which enable one to prove that all solutions have been identified and considered. Which of the solutions should be pursued is only a matter of opinion.

- **Every wicked problem is essentially unique**
  There are no classes of wicked problems such that principles of solutions can be developed to fit all members of the class. Every situation is likely to be one-of-a-kind.

- **Every wicked problem can be considered to be a symptom of another problem**
  Problems can be described as discrepancies between the state of affairs as it is and the state as it ought to be. Removal of the cause poses another problem of which the original problem is a symptom. The higher the level of a problem's formulation, the broader and more general it becomes and the more difficult it becomes to do something about it. The level at which a problem is settled depends upon the problem-solver and cannot be determined on logical grounds. However, one should not try to cure symptoms; one should try to settle the problem at as high a level as possible.

- **A wicked problem can be explained in numerous ways. The choice of explanation determines the nature of the problem's resolution**
  Crimes in the street can be explained by not enough police, by too many

criminals, by inadequate laws, cultural deprivation, too many guns, etc. Everybody picks the explanation which fits his intentions best. What comprises problem-solution for one is problem-generation for another.

Systems design

One of the best existing reports on systems design is made by ***J.P. van Gigch*** (1978). He defines the method as a series of ongoing, cybernetic and fluid design functions. It begins with a question regarding the purpose for the existence of the system and emphasizes the problems in relation to superordinate systems. The search for an alternative design is generally taken beyond the boundaries of the system in question.

This soft system methodology involves ten steps divided into three distinct phases: policy-making/preplanning, evaluation, and action-implementation. It is listed here without the extensive comments belonging to each step as presented in van Gigch's book, *Applied General Systems Theory*. A summarizing block diagram is also presented in Figure 9:1 after the list.

First phase **Policy Making or Preplanning**

Step 1 Problem definition

A. The recipients or clients whose needs are to be met
B. The needs to be met
C. The scope, the extent to which needs will be satisfied
D. The agents involved. All those influenced by the project, considering their interests
E. Evaluation of the agent's world-views according to step 2
F. The methods. Short and general descriptions of methods which will be used to solve the problem
G. The system's boundaries. These should be defined, together with assumptions or constraints affecting the solution or its implementation
H. An enumeration of available resources compared to resources needed
I. A disclaimer to restrict hopes that systems design will provide a solution to everyone's problems

Step 2 Understanding the world-views of clients and planners

A. Premises
B. Assumptions
C. Values
D. Cognitive style

Step 3 Goal setting

A. Needs and wants
B. Expectations and aspiration levels
C. Substitutions, tradeoffs and priorities
D. The morality of systems design (ethical issues)

Step 4 Search for and generation of alternatives

A. Programme alternatives and agents relationships
B. Determination of outcomes
C. Consensus

Second phase **Evaluation**

Step 5 Identification of outputs, attributes, criteria, measurement scales, and models

A. Identification of outputs
B. Identification of attributes and criteria
C. Determination of measurement scale
D. Choice of measurement models
E. Determination of the availability of data

Step 6 Evaluation of alternatives

A. Use of models
B. Measurement of the output of soft systems

Step 7 Choice of alternative

Third phase **Action-Implementation**

Step 8 Implementation

A. Optimizing, suboptimizing, compromising
B. Legitimizing and consensus
C. Experts and expertise

Step 9 Control of systems
Step 10 Evaluation of outputs, auditing and reappraisal

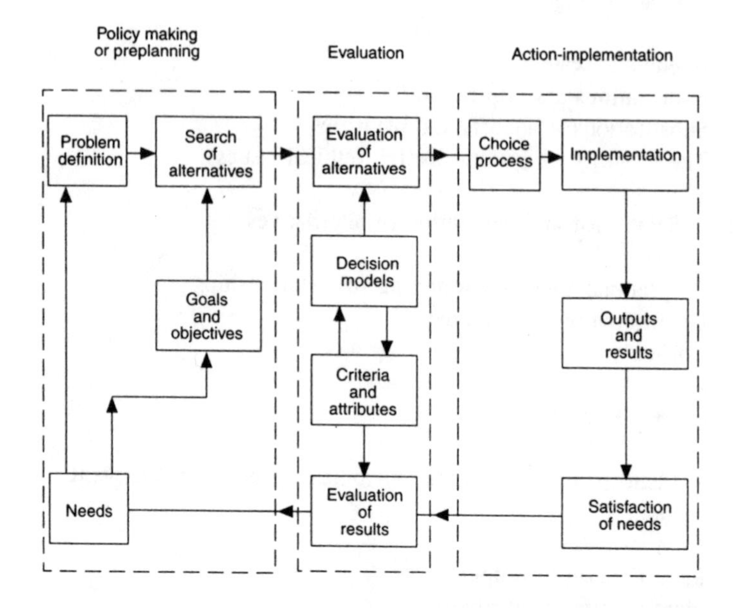

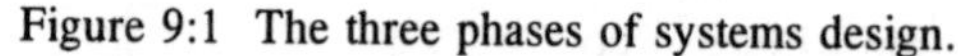
Figure 9:1 The three phases of systems design.

(From J.P. van Gigch, *Applied General Systems Theory*, Harper & Row, 2 Ed. 1978. Reprinted with permission.)

Breakthrough thinking

A new variation of soft systems methodology has been developed by ***Gerald Nadler*** and ***Shozo Hibino***. It was introduced in their book *Breakthrough Thinking* (1990) and is a total, holistic approach to problem solving. The authors state that it combines the best of the visionary and the pragmatic approaches to problem solving and problem prevention. Their concept of the environment in which they have to implement their thinking is free from most illusions. The following quotation from the book gives a good example.

> 'You want to get started with Breakthrough Thinking. I don't give a damn what book you read. All I know or care about is that the billing department overtime is up, costs are high, quality is low, and our mailings are always late. Don't waste my time with fancy theories. I want you to go out there and get the facts on that department. Gather all the information. I want to subdivide this problem into its component parts and analyze the data you come up with. That's the only way that I can make an informed decision.'

Their starting point is that people and organizations need an understanding of the *purposes*, not the problems, in order to move ahead and be successful.

Furthermore, it is even necessary to ask what the purposes are of those purposes. Profit, for example, is only a measure how well a company fulfils a purpose; profit is not the company's mission.

The backbone of the methodology is the seven principles of breakthrough thinking. These should be applied in a coordinated and holistic manner, that is, all the principles should be used simultaneousely rather than consecutively. Furthermore, each principle recurs at a different point in the problem-solving effort. The principles and their application have been summarized from the book in the following arrangement.

The uniqueness principle Whatever the apparent similarities, each problem is unique and requires an approach that deals with its own contextual needs.

1. No two situations are alike.
2. Each problem is embedded in a unique array of related problems.
3. The solution to a problem in one organization will differ from the intended solution to a similar problem in another organization.

The purposes principle Focusing on purposes helps to strip away non-essential aspects in order to avoid working on the wrong problem.

1. Identify stakeholders from whom you want to learn about needs and purposes.
2. Ask questions that expand purposes.
3. Create an array of purposes or intended functions.
4. Set up criteria for selecting the function level.
5. Select the function level.
6. Designate performance measures or objectives for the selected purpose.

The solution-after-next principle (SAN) Innovation can be stimulated and solutions made more effective by working backwards from an ideal target solution.

1. Identify regularities to consider in developing a solution.
2. Develop as many ideas as possible about how you might achieve your selected purpose or a bigger purpose in your goal hierarchy.
3. Organize your ideas into major alternative solutions or systems. Incorporate as many good ideas as you can into each alternative. Each major alternative should be able, on its own, to achieve your purpose and should contain specific strategies for doing so.
4. Add detail and additional ideas to each alternative as needed to ensure its workability and your ability to measure its effectiveness.
5. Select your target solution-after-next by evaluating each major alternative.
6. Try to make your target solution-after-next even more ideal.

7. Develop modifications to the solution-after-next to incorporate the irregularities. Add the details to arrive at a recommended solution.

The systems principle Every problem is part of a larger system. Understanding the elements and dimensions of a system matrix lets you determine in advance the complexities you must incorporate in the implementation of the solution (regarding the systems matrix, see figure 9:2).

1. Assume that the system matrix is empty when you start a project. Start with the purpose of the system and the project.
2. Think in terms of elements first, then expand each element as needed by the dimensions.
3. Transfer any detailing activity from the whole system matrix to individual ones for an element, dimension, or cell if the complexity of the whole becomes too great.
4. Establish the system matrix as a language of communications in networks of like-minded people (meetings, roundtables, etc.).
5. Convert the system matrix into the format used by your organization. A strategic business plan used by a company can be enhanced by system matrix elements and dimensions.
6. Find causes and relationships.
7. Provide an integrating and coordinating framework to handle the many available techniques, tools, and analysis models (cause-effect diagram, statistical control, critical success factors, mathematical programming, chaos theory, optimization modelling, multi-attribute decision analysis, spreadsheets, etc.).
8. Get people to be quality- and productivity-minded in total systems terms.

The limited information collection principle Knowing too much about a problem initially can prevent you from seeing some excellent alternative solutions. This principle serves to focus attention on information that is particularly useful and relevant for the other Breakthrough Thinking principles.

1. Answer questions raised by the development of the solution-after-next for the problem.
2. Use the information and knowledge in the heads of many people doing different work.
3. Ask how an idea or SAN could be made operational.
4. Ask what you would do with the information if it were available in, say, three months.
5. Have a prepared mind, not an empty head.
6. Share information with everyone, not just with an elite coterie.
7. Seek needed information from a wide variety of sources.
8. Study the system matrix of the SAN or recommended solution.

9. Use models and quantitative techniques.
10. Use computer bibliographies, networks, search routines, and databases.
11. Decide what information to collect.

The people design principle The people who will carry out and use a solution must work together in developing the solution with Breakthrough Thinking. The proposed solution should include only the minimal, critical details, so that the users of the solution can have some flexibility in applying it.

1. Try it. Believe in the solution. Talk to people.
2. Hold an informal team meeting during a lunch gathering, etc.
3. Set up a one-time meeting with people who might constitute a good long-term project team.
4. Set up a one-time meeting to plan the problem-solving system using the Breakthrough Thinking principles.
5. Allow for the catharsis of finger-pointing, superficial diagnosis, turf protection and other defensive routines that people are likely to engage in when they first meet on a project.
6. Get the people you involve to be customer-driven and market-oriented through the larger purposes in the array.
7. Involve different people, depending on whether your purpose is to improve an existing system that is in trouble, better a system that is in good shape, or create a new product or service.
8. Include a person or two previously successful at breaking the rules.
9. Include a person or two with a liberal arts bias.
10. Use existing groups. Organizations that have a commitment to teamwork are well positioned to change to Breakthrough Thinking.
11. Keep the energy level high for the Breakthrough Thinking principles, not for initial judging.
12. Seek ways to attain recognition for individuals and groups who have made major breakthroughs.

The betterment timeline principle A sequence of purpose-directed solutions is a bridge to a better future.

1. Incorporate the principle in the overall planning process of the organization rather than treat betterment activities on a separate project-by-project basis.
2. Delegate and decentralize the responsibilities for betterment timeline activities.
3. Use preventive maintenance (PM) for all systems or procedures in the betterment planning. A scheduling system should regularly examine and challenge jobs, units, departments, products, services, policies, etc.
4. Calculate a value-of-change to cost-of-change ratio at the time of decision whether or not a planned change is worthwhile.

Elements \ **Dimensions**	Fundamental: basic or physical characteristics – what, how, where, or who	Values: goals, motivating beliefs, global desires, ethics, moral matters	Measures: performance (criteria, merit and worth factors), objectives (how much, when, rates, performance specifications)	Control: how to evaluate and modify element or system as it operates	Interface: relation of all dimensions to other systems or elements	Future: planned changes and research needs for all dimensions
Purpose: mission, aim, need, primary concern, focus						
Inputs: people, things, information to start the sequence						
Outputs: desired (achieves purpose) + undesired outcomes from sequence						
Sequence: steps for processing inputs, flow, layout, unit operations						
Environment: physical and attitudinal, organization, setting, etc.						
Human agents: skills, personnel, responsibilities, rewards, etc.						
Physical catalysts: equipment, facilities, etc.						
Information aids: books, instructions, etc.						

Figure 9:2 Systems matrix of Breakthrough Thinking.

(Reprinted with permission, © 1990 Gerald Nadler and Shozo Hibino, *Breakthrough Thinking*, Prima Publishing, Rocklin CA).

Systems analysis

The main ideas of systems analysis were originally developed by the RAND Corporation of America. It was created for the study of interaction between science, technology, and society. Several variations have been introduced.

The one presented here is distinctive for the basic concept and has been adopted from ***R. Flood*** and ***E. Carson*** (1988). It is a typical hard systems methodology, consisting of four steps: problem analysis, generation of alternative solutions, evaluation of alternatives, and selection of the optimal alternative. Finally, action is taken based on the selected alternative.

1st step. **Problem analysis** Here the problem and its cost are defined, thus giving an economic measure to use in comparison with alternatives. Two main questions serve as guidelines:

A. What are the limitations of the present system?
B. What is the cost of operating the present system?

Efficient features of the existing system worth retaining are observed.

2nd step. **Generation of alternative solutions** Alternatives to the present system are generated and their main features are examined. Two main questions serve as guidelines:

A. What alternative systems are possible?
B. What would be the operating costs of the alternative systems?

All factors associated with the choice of an alternative, such as the pertinent advantages and disadvantages, should be carefully considered. Comparative economic evaluation between the operating costs of the alternatives should be performed as well as testing of the feasibility of trade-off costs within each alternative.

3rd step. **Evaluation of the alternatives** In this step the capital costs of introducing a new system or improving the present one are assessed. Comparisons between the various alternatives, where both operating and capital costs are taken into account, are made. Two important questions are posed:

A. What are the capital costs of continuing with the present system, and of changing to alternative systems?
B. What comparisons can be made between the various systems, taking all costs into account?

This step includes comparative capital costing, including inquiries into factors such as the rate of return obtained on money invested. Basic principles of capital investment must be considered as inherent in the following questions: What is a reasonable return on the investment and over what time period should the investment be considered? A variety of methods available for calculating the return must thus be considered, along with their advantages and disadvantages. An appropriate method of taking inflation into account in connection with future costs must also be considered.

4th step. **Selection of the optimal alternative** The best alternative is now selected, considering not only economic but also operational, marketing, environmental, and human factors. Two important questions should be asked:

A. What is the most economical solution?
B. Is the most economical solution the best 'all-round' solution?

In these questions lie a real potential for conflict, particularly between quantifiable economic factors and those not so readily quantifiable, such as the quality of human working conditions and environmental impacts (pollution, etc.). Therefore, the best solution is not always the most economically efficient one. A right answer is only right in the limited sense that it reflects the company's objectives at the time of the decision.

Systems engineering

Systems engineering as a 'hard' methodology has its roots in NASA and the early space projects of the 1960s. One of its more generally applicable variations is again presented by Flood and Carson (1988). It has four main phases: systems analysis, systems design, implementation, and operation. The phase of systems design in this case belongs to a typically hard systems methodology. It should therefore not be compared with the same phase in van Gigch's approach presented earlier in this chapter.

Phase 1 **Systems analysis**

A. Recognition and formulation of the problem.
How did the problem arise?
Who are those who believe it to be a problem?
Who decided to implement a planning decision?
Is the problem the right one?
Will it save money?
Is it better to spend the money elsewhere?

B. Organization of the project – A team composed of:
A team leader
Users
Model builders
Designers
Computer programmers
Mathematicians
Economists
Accountants

C. Definition of the system
Breaking down into subsystems and identifying the interactions using flow diagrams representing:
money
energy
materials
information
decisions

D. Definition of the wider system
The role of the system within the wider system of which it is a part, depicted in a flow diagram.

E. Definition of the objectives of the wider system
By use of block diagrams from the system and the wider system, sets of objectives regarding the wider system can be formulated.

F. Definition of the objectives of the system
Initially, these are dictated by the needs of the wider system. Conflicting objectives should be listed and ranked according to their importance. Definitions in economic terms should be used and the efficiency in reaching the objective calculated.

G. Definition of the overall economic criterion
This economic criterion should be directly related to the objectives. Conflicting objectives could be handled by applying a weighting factor to each.

H. Information and data collection
Data-gathering for future modelling of the system and forecasting of the future environment. Use of statistics.

Phase 2 **Systems design**

A. Forecasting

This should be done regarding potential demands, potential activities, and environment for the short, medium, and long terms.

B. Model building and simulation

Predicting the performance in potential operating conditions and real-life environments.

C. Optimization

Identifying the most favourable model performance according to the economic criterion chosen for the study.

D. Control

The solution is checked by different kinds of control loops.

Phase 3 **Implementation**

A. Documentation and approval

A report highlighting the implementation and its critical path should be prepared. This should be approved by all influenced by the planned change.

B. Construction

Creation of hardware and software regarding special control and optimization systems according to a preplanned schedule. Construction of the main system.

Phase 4 **Operation**

A. Initial operation

Enhancing the cooperation between the systems team and the systems users in connection with the delivery of the system. Use of adequate documentation and training of personnel.

B. Retrospective appraisal of the project

Making a report of the whole project. Prospective reoptimization of the project.

GLS simulation

Finally, a hard systems methodology developed by the author will be presented (Skyttner 1988). It has two explicit starting points, one in GLS theory and one in DYNAMO, a computer language developped by ***Jay Forrester*** (see p. ??). The idea is that problems can always be related to one or more of the main flows of matter, energy, or information in a living system. When the problem area is identified it is often found as a consequence of a malfunctioning subsystem. Then measures are constructed and normal values are established for the problematic function. The relationships between this function and other functions or subsystems are then simulated by use of a DYNAMO program and the model is moved through space and time, fed by a continuous series of inputs.

In order to use this idea a complete working methodology has been developed. It includes a bottom-up approach which generally provides good knowledge of functional principles as well as more detailed answers. The methodology includes the following steps:

Selecting the perspective

A. Problem definitions
B. Construction of relevant measures
C. Problem selection
D. System-boundary definition
E. System-level definition
F. Construction of system block diagram

Identifying system processes

A. Identification of essential subsystems
B. Identification of matter/energy, information flows
C. Construction of causal loops for model subsystem
D. Construction of interaction matrix
E. Construction of schematic description

Converting to Dynamo model

A. Identification of useful model subsystems
B. Construction of causal loops for useful model subsystems
C. Construction of schematic charts for model subsystems
D. Construction of Dynamo flow charts
E. Definition of quantities and processes
G. Converting step d) and e) to Dynamo computer language

Running the Dynamo model

A. Simulating by changing the parameters
B. Verifying the model
C. Validating the model
D. Documentation of the work

The methodology has, for instance, been used to rationalize the operation of the Swedish search and rescue (SAR) system at sea. The main problem was that there was no specific knowledge of what could be considered to be critical functions or parameters influencing the overall system efficiency. Examples to illustrate some of the working steps are taken from this mission.

The overall measures constructed were *human survivability*, *economical salvage,* and *ecological recovery*. The selected problem was to examine the timer function which was estimated to be critical for the system.

The system boundary was taken to be the geographical area at sea that was the responsibility of the Swedish SAR system. This also included certain land-based locations, such as rescue centres, radio stations, etc. The defined boundary might also be temporarily augmented or changed through cooperative agreements with other national systems.

For practical purposes, the system was defined as belonging to the organizational level. This does not exclude the possibility that it sometimes works on the supranational level.

An extract from the identification of the twenty essential subsystems is shown in the table below (in principle, this is the same kind of identification that was done for a supermarket on page 83).

Matter/energy and information

1.	Boundary	A.	Responsibility area limit
		B.	Air/water-vehicle range limit
		C.	Radio/radar/look-out range limit
2.	Reproducer	A.	Renewed international agreements
		B.	The need for better systems
		C.	Commercial interests
		D.	Recruiting officers

Matter/energy

3.	Ingestor	A.	Oil
		B.	Electricity
		C.	Water
		D.	Ventilation
		E.	Food
		F.	Ships, ambulances, aircraft
		G.	Supplies, spare parts
		H.	People
4.	Distributor	A.	Aircraft pilots, ship's crew, ambulance drivers, servicemen
		B.	The sea, the air
		C.	Pipelines, ventilation shafts
		D.	Cables
5.	Converter	A.	Oil converters for climate/motion
		B.	Electricity converter for light, radio, radar, telex
		C.	Water coolers

D. Human living conditions
E. Food preparation and consumption

The extract which follows below is from the identification of matter/energy and information flows between the different subsystems. The alphanumeric designations represent the various subsystems with their subdivisions. It is taken from the list above where for example a flow exists between the oil inlet 3A and the oil tank 7A (not shown in the above list: the complete list of all subsystems in a living organism with appropriate numbers is given on page 78).

Matter/energy input	From	To
Fuel	3A	7A
	7A	5A
	5A	9B
	9B	8D, 8E
Electricity	3B	7B
	7B	5B
	5B	9B
	9B	8F
Water	3C	7C
	7C	5C
	5C	9B
	9B	8E, 8F
Air	3D	7D
	7D	5D
	5D	9A
	9A	8F
Food	3E	7E
	7E	5E

Note that three main system flows can always be identified in living systems, those of matter, energy, and information. All flows are entering the system and are to some extent stored. Inside the system, information processes regulate them. This is accomplished by continuously sensing the system's status. After feeding the processes within the system and simultaneously transforming their own content, the flows leave the system.

The list identifying matter/energy and information flows is transformed into a causal loop diagram according to figure 9:3.

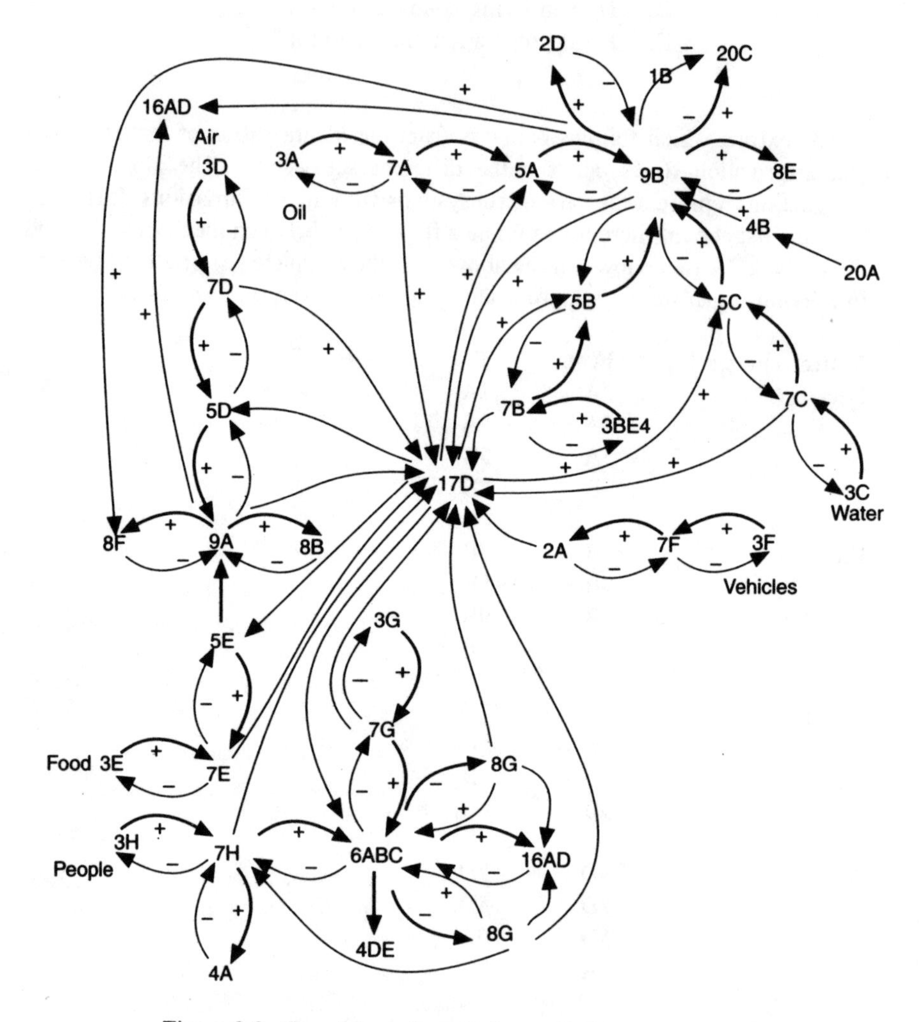

Figure 9:3 Causal loop diagram for matter/energy flows.

The subsequent step was to construct an interaction matrix where all essential subsystems with their subsections were contained in the rows and columns. Listings of the system flows, together with the causal loops, were used to indicate negative and positive interaction.

The schematic description was then made with the main purpose of enhancing the total understanding of the system. Jointly with the causal loops and the interaction matrix, this step completes the symbolic presentation of the system.

The conversion to a DYNAMO model was done by transforming the matrix into Forrester schematics with the help of a special algorithm (J. Burns 1977). The strategy of this algorithm is to embed Forrester's simulation methodology in a mathematical framework which specifies concrete guidelines and rules when a computer program is written. The use of the algorithm ends up with the identification of minor subsystems and a classification of quantities for the Forrester schematics. These data constitute building parameters for the DYNAMO flow chart. Based on this, quantities and processes were defined and the DYNAMO computer program was written. Thanks to the many detailed preceding steps and the ingenious algorithm by Burns, the computer program was very easy to write.

As a result of the simulation, the general knowledge of functional principles for the SAR organization was refined. In the system investigated a three-part M/E process represented the ultimate system goal – to save human lives. This was done by

- removing persons from the danger area, and
- to make this possible, sometimes doing construction and repair work, and
- if necessary giving medical treatment.

All this was centred around the motor, producer, and storage functions and was highly time-dependent. The timer function affects such subsystems as channel and network as a result of current radio communication conditions (e.g. effects of sun spot cycles, sun flares, metereological phenomena such as thunderstorms, etc.). Also Motor, Distributor, and Storage are directly timer dependent as a result of seasonal and diurnal variation, which in turn affects the rescue and survival time.

Method versus problem

In this chapter only the main steps of the methodologies were dealt with. This is because the appropriate sub-steps often include the use of sophisticated statistical, mathematical and economic tools and models, discussion of which is outside the scope of this book.

Of the many systems methodologies available, some clearly work well in certain situations but not in others. Naturally, each has its strengths and inevitable limitations. As a rough guide on how to choose the most suitable, the division between hard methodologies and soft methodologies made by Peter Checkland (1981) is a useful starting point.

Soft methodologies are used in the absence of a concrete definition of ill-structured problems in which objectives and purposes are themselves often problematic. Such problems mainly exist in connection with various kinds of social systems and cannot be given an exact form or be forced into a predetermined structure. They are multivariate problems to be dissolved rather than solved. Soft methodologies are characterized by mutual understanding and the generation of learning. Typical soft system methodologies presented here are Systems design and Breakthrough thinking.

Hard methodologies are goal-oriented in the solution of structured problems where well-defined objectives and constraints exist. Normally these methods use some kind of means-end analysis. They work according to the implicit belief that any problem can be solved by setting objectives and finding optimal satisfying alternatives directed towards a defined problem solution. Identifying, designing, and implementing are the main phases of this approach. Systems analysis, systems engineering, and GLS simulation have been chosen as examples here.

Finally, from the plethora of existing systems methodologies the following are well known and therefore deserve to be mentioned, even though they are not considered here, as they are more specialized.

- Checkland's own Soft Systems Methodology, SSM (Checkland 1981)
- Total Systems Intervention, TSI (Flood & Jackson 1991)
- Strategic Assumption Surfacing and Testing, SAST (Mason & Mitroff 1981)
- Critical Systems Heuristics, CSH (Ulrich 1983)
- Scenario techniques (von Reibnitz 1988)

An overall examination of the existing systems methodologies indicates that most of them are partitioned into about a half a dozen steps, while some of them distinguish up to ten steps. The steps and phases are sometimes mixed, both logically and chronologically. ***Kjell Samuelson*** (1976) has compiled a list of 45 logical/chronological steps that can be distinguished in various methodologies, although they sometimes overlap. It has the following form.

1. Ultimate objectives
2. Goals statement
3. Constraints
4. Needs and requirements
5. Definition of project or problem
6. Idea and inception
7. Conception (or conceiving of solutions)
8. Conceptual design with functional model
9. Feasibility study (impact declaration)
10. General systems proposal and masterplan
11. Object system analysis (and description)
12. Observation of existing systems
13. Alternative designs outlined
14. Identifying alternative technologies
15. Preliminaries (survey, analysis, design)

16. Design evaluation
17. Specifications
18. Information system macrostructuring
19. Information study (e.g. precedence analysis)
20. Processing study
21. Details study
22. Decision analysis
23. Data analysis
24. Data construction and microstructuring
25. File structuring
26. Component design
27. Simulation
28. Conversion
29. Pilot system (experiment)
30. Pilot evaluation (empirical testing)
31. Selection of system
32. Implementing (and programming)
33. Installation
34. Testing
35. Evaluation
36. Operation
37. Training
38. Maintenance
39. Organizing production resources
40. Management of production resources
41. Product distribution
42. Disposition of waste
43. Result reporting and follow-up review
44. Redesign
45. Cessation

Review questions and problems

1. What is the main difference between Systems design and Systems improvement?

2. Discuss why solutions to social problems tend to induce problems which are more severe than those which they were intended to solve.

3. Breakthrough thinking as a methodology states that it is necessary to ask what the purpose is of an organization's purpose. Does this make sense?

4. The methodologies in this chapter are designed to solve large-scale, intertwined problems of complex systems. Does anything prevent their application to small organizations?

5. What is the difference between solving a problem compared with resolving it?

6. If use of hard systems methodologies is said to produce a defined problem solution, what is then the aim of soft systems methodologies?

7. Systems science states that it gives special attention to the dangers of technological fixes and mega science. Is this a justified statement?

10 The Future of Systems Theory

- **Science of today**
- **The world we live in**
- **The need for change**
- **A new paradigm**
- **How to write the instruction manual**

'... one outstandingly important fact regarding Spaceship Earth, and that is that no instruction book came with it.' (Fuller 1970)

In the foregoing chapters, an attempt has been made to present the current state of systems theory and some of its applications. To predict the future of an area which still has not celebrated its fiftieth anniversary is of course extremely difficult. Any effort to predict the coming fate of systems theory must however be made from two starting points. One with respect to the state of science in general where its environment is to be found and the other with regard to the state of the world we live in where it should be used. After all, the highest purpose for the intellect is the search for general principles that would allow us better to understand, predict and manage the world's problems. We begin by taking a critical look at the science of today.

Science of today

In the summary of the scientific development in Chapter 1, it was possible to see how the search for *the general* involved studying *the particular*. The general has however remained limited, often hampered by a sceptical attitude toward holism. Taking ourselves as the starting point of all thinking we have succeeded in being unaware of much that is controlled by the surrounding context. Traditional European individual-centred heritage has held man to be exceptional and superior, the very owner of nature. The supremacy of thought and reason, of cause and effect, as a guiding star for the perfect rational man is still held as an ideal. Apparently, we have a long-standing fear that rationality will be overwhelmed by chaos and the spiritual by the sensual.

The same can be said of dualism or polarity, the traditional Western way to arrange the world and life in mutually exclusive concepts. We think in terms of either/or, black and white, good and evil, defining things by their opposites.

Society still holds specialization as natural, inevitable and desirable, even though man represents the least specialized creature on earth. The earlier scientific and technological view of nature as a grand mechanistic machine with no intrinsic values persists. So do the traditionally positivistic attitudes based on Frances Bacon's 16th century ideas on the extraction of maximum benefit from nature.

However, the worst thing is that the established scientific community has such a strong resistance to change, fortified by deeply rooted private interests. These interests include military-industrial enterprises, oversized weapon bureaucracies, influential secret weapon laboratories, universities with miltary research grants, elitist expert groups trying to control the arms race, and of course personal patents rights. To these can be added a customary resistance to change from an uninformed general public, from the unions which oppose the disappearance of jobs and from the politicians that they support.

Together, these convictions continue to encourage control, exploitation and destruction of nature through scientific 'force'. Short-term profit is still gained through neglect of the second law of thermodynamics. The bill for these illusory benefits will however have to be paid. Some of the consequences are already clearly visible and have resulted in the entropy of global pollution and the collapse of nature.

Modern cross-scientific research, which is growing in popularity, does not change the situation. To place more and more specialized areas side by side under the same thematic roof is inadequate, so long as the involved disciplines depend upon their own methods and language.

It is therefore quite understandable that today's science and technology often give rise to a deep distrust of current research, especially among the younger generation. An activity which resulted in the development, production and stockpiling of the means to kill ever-more people at ever-greater distances in ever-shorter time promotes a general scepticism. The same distrust exists with regard to civil nuclear science and technology. Three Mile Island and Chernobyl are by no means forgotten.

The world we live in

History has always witnessed self-destructive individuals and societies. The dilemma of our generation is that destructive power has gained such devastating strength. Not so long ago the human race was small and relatively powerless and its actions could not significantly affect the grandeur of Nature. Today, human beings have access to power which threatens their own habitat and existence. Modern industrial societies are so intertwined that the consequences of bad decisions, harmful technologies and self-destructive behaviour will be felt across all traditional boundaries.

Now, at the end of the 20th century, man has taken over many of the control mechanisms of the global system. In fact, today he is part of the mechanism itself and problems earlier managed by nature are now a human responsibility. In turn the problems facing human societies have increased in complexity and the stakes are set higher. All too often mistakes cannot be corrected and the defrayal of errors is impracticable. Very often short-range solutions have shown themselves to be both risky and highly uneconomical. Also, the once second-order effects – such as exhaust gases from motor vehicles – have become primary problems influencing the global climate.

We have created for ourselves a world of systems, existing in a state of fragile stability rather than natural balance, and demanding constant maintenance. Many of these systems are both wasteful and dangerous. During the 1980s we witnessed an increasing number of ecological, social and technological disasters. From an inventory of the 20th century's most serious threats to life the following stand out in our memory.

- A nuclear war would pose the most extreme threat; the possible consequences are convincingly analyzed (nuclear winter, etc.). Since the Cuba crisis at the brink of such a war in the 1960s, the risk seems to have diminished substantially. After the breakdown of the Soviet Empire and the dismantling of the cold war, the so called 'holocaust clock' in Washington D.C. has been set back more than an hour.

- The ecological catastrophe, of which local overpopulation is the major imminent threat, relates reciprocally to most of the other problems occurring throughout our environment. A significant change in the global climate and a diminishing ozone layer are consequences of the activities of the inhabitants of an inequitable world, seeking an ever-better life. The same goes for food and water pollution and the spread of deadly diseases such as cancer and AIDS.

- Social and economical degradation can be illustrated with pathological examples from the former communist countries or in the Middle East, where there are signs of the devastating potential of modern ideological and religious fanaticism. The relatively rich and relatively stable Western countries are however no exception: bureaucratic paralysis, alienation, criminality and political corruption are producing degenerative, even fatal effects.

- Great planning disasters where the building of the Aswan Dam on the Nile River is the most spectacular. It is a perfect example of the impact of man-made systems on natural systems, initiated by political prestige and realized as a demonstration of power. The motivation of the dam was a solution of the age-old problem of the annual flooding of the Nile. After the conclusion of the project several serious new problems arose. In the eastern Mediterranean the food chain was broken, thereby severely shrinking the fishing industry. Erosion of the Nile delta took place, causing soil salinity in upper Egypt. In the absence of annual dryness, the water-borne snail parasite Bilharzia grew explosively, initiating an epidemic of intestinal disease along the Nile. All these side-effects were never considered by the responsible leaders of the project.

- Technological breakdown is the most spectacular threat to modern man, sometimes killing thousands of people. Such breakdowns are presented by the media as front-page stories but are seldom given an adequate background analysis. The strong correlation between social degradation and technical disasters is obvious: even first-class passengers in the best of existing worlds die when arrogance is the managerial lodestar of their Titanic. One example fresh in our memory is the exploding spacecraft Challenger spreading its burning wreckage off the coast of Florida while being watched by millions on television screens around the world.

In 1989 an Iranian aircraft was gunned down by an apprehensive crew aboard an American cruiser in the Persian Gulf. One year later a bomb was hidden on board an American aircraft; it exploded over Lockerby and more than 500 people lost their lives. A similar tragedy, that of the Korean aircraft blown up over the Kuriles by an over-zealous Soviet jet-fighter pilot, took the lives of nearly 300 people.

Traffic on the seas offers other examples. An English ferry departed from a Dutch port with its stern gate open, took in water and suddenly sank, taking with it more than 400 passengers. When a Swedish passenger ferry was set alight by a pyromaniac while at sea, the loss of life exceeded 150. An overloaded passenger ship was hit by a storm in Malaysian waters and went to the bottom with all of the more than 3500 people on board – the world's largest peace-time ship disaster.

Being on land can be just as disastrous. One of the most horrifying examples is the escape of poisonous gas at a factory in Bhopal, where more than 300 persons were gassed to death. In a similar accident in New Mexico in 1991 gas in culverts under the street exploded, destroying a whole main street and also killing several hundred people. Another gas catastrophe occurred in the former USSR when a crowded train ran into a cloud of gas leaking from a tube running parallel to the railway. A spark from the train caused an explosion, which devastated a huge area and killed more than 600 persons.

- Social disasters have always been plentiful. One typical example is the 1992 riot in Los Angeles with many dead and the destruction of property running to billions of dollars.

- The Chernobyl disaster, in which a nuclear plant melted down with immediate, long-term and still unpredictable consequences, represents the great number of still current, and potential, ecological, social and technical catastrophes.

To the sad list presented above can now be added the 900 more recent victims who lost their lives when a Swedish passenger ferry abruptly sank in stormy Baltic waters. According to new sea-safety regulations it had no radio officer on board, and its two automatic satellite-operated emergency radio beacons were never activated during the catastrophe. After this disaster some kind of 'distrust movement' showed itself in the daily press. Don't trust technology and don't trust those who trust it - especially if it concerns nuclear plants, super-ferries, jumbojets, or spaceships. Many people express their private concern that anything whatsoever could strike them at any time - a fear related to the that of mediaeval man waiting for the Last Judgement. It would be a great mistake not to take this concern seriously and not to admit its justification.

The need for change

Our present world-view is the result of a 400-year-old scientific project. We have travelled to the moon, split the atom, succeded in transplantation of hearts and rebuilt genes. Nevertheless, we are not satisfied with the outcome of the project. Apparently, its methods do not coincide with today's problems - complexes of interrelated processes on multiple levels - which are characterized by a general air of being insoluble.

Both ordinary people and scientists feel that science - and its offspring technology - no longer enhance the quality of their lives, but are in fact systematically reducing it. Even obstinate economists have begun to realize that national figures of growth only reflect illusory economical progress. Somehow important qualities like judgement, sense for proportion, respect and responsibility are missing. With problems relating to the whole domain of human knowledge, from philosophy to cellular biology, solutions have to be based on something more than the old scientific paradigm.

Positivism, lacking in foresight and comprehensive views, now gives a diminishing return in area after area from social science to quantum physics. Already in 1960 the well-known management scientist Russel Ackoff lamented: 'We must stop acting as though nature were organized into disciplines in the same way that universities are.'

After the end of the cold-war era (1945-90), tendencies towards disintegration have grown strong in many former communist communities as well as in some of the Western capitalist societies. No longer distracted by the cold war, the general impact of overpopulation, energy shortages, environmental pollution, organized crime, deforestation, climatological deterioration, civil wars and global inflation has become visible, giving rise to new pressures on governments and planners.

It is likely that the planet will meet serious instabilities in its natural, social and economic systems over the next fifty years. A collapse seems even probable when the closely interlinked system parameters of time, consumption and population are examined and related to each other (see for example Forrester's *World Dynamics*, 1974).

Accelerated technological innovations are no longer a realistic solution because the cost of developing new control systems to control the adverse impact of old ones rises exponentially. Moreover, systems which have been neglected for a long time have already been irreversibly changed. The traditional Western business-as-usual policies will come to an inevitable halt with deteriorating weather conditions, deforestation, desertification and the extinction of plants, birds, fish and other animals. Contaminated oceans, seas, rivers, soils and pertinent health problems with decreasing life expectations will bring about a very uncertain future.

To these problems must be added the impact of growing global unemployment – a phenomenon originating from the combined effects of overpopulation and automation. This will rapidly increase the breach between both citizens and countries and create hostile reactions, especially against the rich western area. A consequence will be an immigration pressure, already clearly visible in both Europe and United States. Economists have claculated that, to reduce global unemployment, there is the need for one milliard new jobs within a five-year period. This is more than all jobs existing today in the industrial countries taken together and a completely unattainable goal.

The relation between mankind and the large-scale technological systems seems to be of a dubious kind, something concluded in the previous section. The common characteristic of the examples given was the unpredictable break-down of these systems. What is remarkable is that the unpredictability is experienced by those outside the system more than by those inside it. From the outside it seems reasonable to think: why should Chernobyl *not* break down, with its corrupt management, primitive technical solutions and poorly trained personnel?

Those on the inside ultimately responsible for the disasters are as always technocrats, severely lacking in an imaginative ability to systematize the consequences of malfunctions. New insights into the design of more sophisticated humane systems as well as the redesign of earlier manual systems are high on the agenda.

The world to be lived in is also waiting for a better relationship between man and his environment. From the study of pollution, of the destruction of natural resources and of the ecological balance, has evolved an expectation of something new. More and more, the whole earth is being seen as one and, in a sense, as alive. A view is emerging where each individual is regarded only as a part of an organized wholeness greater than himself. Our environment is becoming a sphere no longer separated from human action, ambitions and needs. Kenneth Boulding says the following with reference to a high-cost prestigious project such as the building of a huge dam (possibly Aswan):

'There are benefits of course, which may be countable, but
Which have a tendency to fall into the pockets of the rich,
While the costs are apt to fall upon the shoulders of the poor.
So cost-benefit analysis is nearly always sure
To justify the building of a solid concrete fact,
While the ecologic truth is left behind in the abstract.'
(from *A Ballad of Ecological Awareness* 1973)

Today global problems are seldom associated with lack of awareness and knowledge – instead they regard questions of will and world-view.

A new paradigm

A basic distinction of science is that is it uses observations, measurements and experiments to answer questions associated with problems. But new and relevant questions must be posed if one is to obtain new and useful answers. The heterogeneous groups of systems theorists are held together by the predilection for what they see as new ideas and a new way of thinking. Many of them have a relativistic attitude regarding the modern world-view with its instruments and procedures (Western science and technology). Our present world-view is seen as one among many conceivable, and probably not providing the most desirable course for humanity.

Systems theorists see common principles by which systems of all sizes operate and they strive for an interdisciplinary science with a body of comprehensive knowledge, attempting for a universal application. A common language and area of concepts, to be used by collaborating researchers, is seen as the prime means to overcome the fragmentation of knowledge and the isolation of specialists.

Other researchers state that there are no other ways to solve human problems than by traditional science. Relativism is said to lead anywhere and nowhere at all. They see no special systemic rules and human creativity needs to go for invention more than for discovery. The world needs concrete, immediate technical solutions of problems like AIDS, cancer and traffic accidents and not general principles.

Systems theory was born in the pre World War II optimism, in an era of increasing resources, as an alternative answer to needs which were then considered pressing. Now, fifty years later, with a protracted international recession and decreasing resources, the situation has changed. The established and well-entrenched academic world seems to have too much to lose in accepting a new perspective and therefore regards a new paradigm as a threat. Several of the well established and still reductionist academic disciplines of today seem to have forced relations to systems science. Furthermore, in the eyes of unsuspecting members of academia, systems theory and systems science has always been a subset of computer science. This misconception has often been painstakingly cherished by computer scientists in order to lay hands upon existing but diminishing funds.

As a university discipline, systems science has never striven for an educational monopoly, a fact that has probably contributed to its decline. It is in the nature of systems science that other areas can cross its indistinct boundaries and use its methods. Its territory does not comprise a specific area of empirical reality and its specific methods are no more specific than the fact that they are used by other disciplines. Finally, large-scale systems thinking in the shape of social engineering supported by extensive computerization of the citizen's everyday life has given systems theory an unmerited bad reputation.

Taking this last-mentioned fact as a starting point, one of the most aggressive critics of the systems movement (and also one of the most zealous) has been *Robert Lilienfeld*. In his book *The Rise of Systems Theory* (1978) he conducts a general attack on its whole spectrum of ideas and methods. From his critiques the following main points deserve to be mentioned:

- Systems theory is the latest attempt to create a universal myth based on the prestige of science. Earlier myths has been based on theology or philosophy.

- Systems thinkers have a special weakness for definitions, conceptualizations and programmatic statements, all of a vaguely benevolent moralizing nature, without concrete or even scientific substance. Rather arbitrary normative judgements are built into its technical apparatus.

- In the eyes of the 'universality' of systems theory all things are systems by virtue of ignoring the specific, the concrete and the substantive.

- As as theory, systems philosophy is a mixture of speculation and empirical data, neither of them satisfactory. It is an attempt to stretch a set of concepts into metaphysics that extends beyond and above all substantive areas.

- Systems theory is a theory with applications which has never been really tested.

- The systems view of society as an organism appears attractive to intellectuals, who will see themselves as the brain and nerve centre of the organism, dealing as they do with symbolic and conceptual matters.

- Systems theory is not a genuine philosophy and is not a science; it is an *ideology* and must be considered as such. As an ideology it promotes *meritocracy*: freedom to command for those at the top of the hierarchy and freedom to obey for those locked into the system. Its authoritarian potential seems striking to all but the systems theorists themselves.

Lilienfeld does not appear to understand that systems theory can help us to explain why an omnipresent nomenclature is unable to let people alone. The systems theorist knows that radical intervention in natural and social systems is a certain way to achieve surprising effects or to initiate a breakdown. He also knows that the solution of one problem often creates a new, more serious one. Systems scientists are not social engineers, but on the other hand they are very capable of explaining why that discipline also often fails.

For the practitioners of systems theory, Lilienfeld's declaration that the area has ideological overtones makes no sense. On the other hand, it is quite obvious that the systems movement embraces certain ethical dimensions. These were reactivated as a necessary response when humanity seemed to approach nuclear extinction during the most intense cold-war era.

The systemic challenge to the traditional academic world has unfortunately been taken as a threat against its existence. The scientific and educational impact of systems theory has therefore been opposed as an alternative to the old scientific paradigm. Consequently, systems theory and systems science have lost much of their earlier popularity and it is no longer possible to study those areas as independent subjects in Swedish Universities. Perhaps the Scandinavian outlook is too short-sighted, but a general impression is that this diminishing popularity is an universal trend.

At the time of writing, the future of Systems Theory seems bleak. Its underlying principles may still be neglected for some years, but the growing number of international crises will force the establishment to resort to all means, including systems science. To translate message from mother Gaia and to navigate Spaceship Earth is after all a completely holistic science.

How to write the instruction manual

Buckminster Fuller was one of the first proponents for an operating manual for Spaceship Earth. His own life and work was an attempt to contribute to this symbolic idea which can be followed in his books. This concept does not deal with gradually increasing improvements of the present situation. Instead it deals with a total disengagement from old notions and traditions in order to create new opportunities.

In reality our old world-view has not been kept up to date to take account of contemporary change. The metaphor of Spaceship Earth, however, fulfils the need for an up-to-date model of our world. A spaceship is a closed system with well-defined and discernible rules for the survival of its crew. Stores and waste irrevocably set the limits for the ship and its living system.

Those skilful in science and technology seem no more adept to navigate the ship than practitioners of other knowledge areas. All things considered, the great innovations of the 19th and early 20th centuries are responsible for most of our problems – something that technology itself is currently trying to find solutions for. The objective, value-free position of science has erroneously been taken for good science. After two world wars and a trip to the brink of a nuclear holocaust, humanity has probably learnt that science does not *per se* convey a survival advantage. As a realizer of human interests, science has to be founded on transcendental values that exist *outside* itself. A science that cannot address the values, needs and questioning of modern life and inspire its support, will soon lose its public justification.

A basic part of the manual will instruct science, technology and all their institutions to serve man, not man to serve them. It will prove that the technological imperative – that every technology which *can* be developed also *ought* be developed – has long been both obsolete and harmful. Probably, it will introduce its own variation of Heisenberg's uncertainty principle. This will tell us that if we decide on an excessive standard of living we have to choose a low quality of life. If, on the other hand, we want a better quality of life, we have to give up an excessive standard of living.

Regarding possible methods it will say that old methods must now be balanced by the use of new methods, taken from all human knowledge areas including music, art and philosophy. The results should be compiled when complementing each other – if incompatible, further methods and more analysis should be pursued.

Unfortunately, this manual does not yet exist and it is doubtful whether it will be written in the future. It is, however, the belief of the author that some of the ideas presented in this book should well qualify for inclusion in the manual.

Review questions and problems

1. The European cultural tradition has held man as owner of nature while other cultures see man as a part of it. What has caused this difference in rules of conduct?

2. Why does cross-scientific research not cooperate in the unification of science?

3. Give some examples of second order effects produced by modern technology which have now become first order problems.

4. Give an example of a great planning disaster planned and implemented by engineers and opposed by a large local population.

5. What is the main difference between the old scientific paradigm and the systems paradigm?

6. Which of the seven main points containing Lilienfeld's criticisms of systems theory can be considered the most serious?

7. What should be the main reason for the diminishing popularity of systems theory?

References

Ackoff, R. (1960) *Systems, Organisations and Interdisciplinary Research*, General Systems Yearbook
Ackoff, R. (1970) *A Concept of Corporate Planning*, John Wiley, NY
Ackoff, R. (1971) 'Towards a System of Systems Concepts', *Management Science*, Vol. 17, No. 11, Providence, RI
Ackoff, R.(1981) *Creating the Corporate Future*, John Wiley, NY
Ashby, R. (1964) *An Introduction to Cybernetics*, Chapman & Hall, London
Aulin, A. and Ahmavaara, A. (1979) 'The Law of Requisite Hierarchy' *Kybernetes*, No. 8, London
Austin, J. (1978) *Chance and Creativity: The lucky Art of Novelty*, Columbia UP, NY
Beer, S. (1972) *Brain of the Firm*, Penguin Press, London
Beer, S. (1979) *The Heart of Enterprise*, John Wiley, NY
Bergström. M. (1990) *Hjärnans resurser*, Seminarium Förlag AB, Jönköping
von Bertalanffy, L. (1955) 'General Systems Theory', *Main Currents in Modern Thought*, 71, 75, New Rochelle, NY
von Bertalanffy, L. (1967) *Robots, Men, and Minds*, NY
Bidgoli, H. (1989) *Decision Support Systems*, McGraw-Hill, NY
Boulding, K. (1956) 'General Systems Theory - The Skeleton of Science' *Management Science*, 2, Providence, RI
Boulding, K. (1964) 'General Systems as a Point of View' in Mesarovic's *Views on General Systems Theory*, John Wiley, NY
Boulding, K. (1978) *Ecodynamics*, Sage Publications, London
Boulding, K. (1985) *The World as a Total System*, Sage Publications, London
Bowler, D. (1981) *General Systems Thinking*, North Holland, NY
Burns, J. (1977) 'Converting Signed Digraphs to Forrester Schematics', IEEE, Vol. SMC-7, 10
Campbell, N. (1953) *What is Science?*, Dover Publications, NY
Checkland, P. (1981) *Systems Thinking, Systems Practice*, John Wiley & Sons, Chichester
Cherry, C. (1966) *On human communication*, MIT Press, London
Churchman, W. (1979) *The Design of Inquiring Systems: Basic concepts of Systems and Organizations*, Basic Books, NY
Coakley, T. (1991) *Command and Control for War and Peace*, National Defense University Press, Washington DC
Cook, N. (1980) *Stability and Flexibility. An Analysis of Natural Systems*, Pergamon Press, NY
Deutsch, K. (1966) *The Nerves of Government*, The Free Press, NY
Dörner, D. (1980) 'On the Problems People have in Dealing with Complexity' *Simulation and Games*, 1, Beverley Hills, CA

Dreyfus, H. and Dreyfus, S. (1986) *Mind over Machine*, Basil Blackwell, NY
Edelman, G. (1991) in M Benedikt's *Cyberspace*, MIT Press, Cambridge, Mass.
Einstein, A. (1921) *Geometrie und Erfahrung*, Sitzungsberichte der Preussichen Akademie der Wissenschaft, Verlag der Akademie der Wissenschaften, Berlin
Fivaz, R. (1989) *L'ordre et la volupté*, Presses polytechniques romandes, Lausanne
Flood, R. and Carson, E. (1988) *Dealing with Complexity*, Plenum Press, NY
Flood, R. and Jackson, M. (1991) *Creative Problem Solving*, John Wiley, NY
Forrester, J.W. (1969) *Urban Dynamics*, MIT Press, Cambridge, Mass.
Forrester, J.W. (1971) *World Dynamics*, Wright Allen, Cambridge, Mass.
Fuller, B. (1970) *Operating Manual for Spaceship Earth* Carbondale Ill, Southern Illinois University Press
Fuller, B. (1992) *Cosmography*, Macmillan, NY
van Gigch, J. (1978) *Applied General Systems Theory*, 2nd Ed., Harper & Row, NY
van Gigch, J. (1991) *System Design Modeling and Metamodeling*, Plenum, NY
von Glaserfeld, E. (1987) 'The Construction of Knowledge: Contributions to Conceptual Semantics' Intersystems Seaside, Ca.
Hillis, D. (1985) *The Connection Machine*, MIT Press, Cambridge, Mass.
Holling, C. (1977) *New Directions* in *Futures Research*, eds Linstone and Simmonds, Addison-Wesley, Reading, Mass.
de Jardin, T. (1947) *The Phenomenon of Man*, Harper & Row, NY.
Jaynes, J. (1982) *The Origin of Consciousness in the Breakdown of the Bicameral Mind*, Houghton Mifflin, Boston, Mass.
Jordan, J (1968) *Themes in Speculative Psychology*, Tavistock Publications, London
Klir, G. (1985) *Architecture of Systems Problem Solving*, Plenum, NY
Klir, G. (1991) *Facets of Systems Science*, Plenum, NY
Koestler, A. (1967) *The Ghost in the Machine*, Arkana, London
Kroenke, D. (1989) *Management Information Systems*, McGraw-Hill, NY
Langefors, B. (1993) *Theoretical Analysis of Information Systems*, Studentlitt., Lund
Langton, C. (1989) *Artificial Life*, Santa Fé Institute Studies in the Science of Complexity, Vol. 6, Addison-Wesley, Redwood City
Laszlo, E. (1972) *Introduction to Systems Philosophy*, Harper & Row, NY
Laszlo, E. (1972) *The World System*, George Braziller Inc., NY
Lawson, J. (1978) 'A Unified Theory of Command and Control', 41st Military Operations Research Symposium
Libet, B (1985) 'Unconscious Cerebral Initiative and the Role of Conscious Will in Voluntary Action', *Behavioral and Brain Sciences*, 8, Cambridge, Mass.
Lilienfeld, R. (1978) *The Rise of Systems Theory*, John Wiley, NY

Litterer, J. (1969) *Organisations: Systems, Control and Adaption*, John Wiley, NY
Lovelock, J. (1988) *The Ages of Gaia*, Norton & Co, NY (reissued OUP, Oxford 1995)
Lovelock, J. (1979) *Gaia: A New Look at Life on Earth*, Oxford University Press, Oxford (reissued 1995)
MacKay, D. (1969) *Information, Mechanism and Meaning*, MIT Press, London
MacLean , P. (1973) *A Triune Concept of the Brain and Behavior*, University of Toronto Press, Toronto
Mason, R. and Mitroff, I (1981) *Challenging Strategic Planning Assumptions*, John Wiley, NY
Maturana, H. and Varela, V. (1992) *The Tree of Knowledge*, Shambala, London
Miller, J. (1978) *Living Systems*, McGraw-Hill, NY
Miller, J. (1990) 'Introduction: The Nature of Living Systems' *Behavioral Science*, Vol. 35, 3, La Jolla, Ca
Nadler, G. and Hibino, S. (1990) *Breakthrough Thinking*, Prima Publishing Rocklin, Ca
Namilov, V. (1981) *Faces of Science*, iSi Press, Philadelphia
Nörretranders, T. (1993) Märk världen Bonnier Alba, Stockholm
Powers, W.T. (1973) *Behavior: The Control of Perception*, Aldine de Gruyter, Hawthorne, NY
Pribram, K. (1969) 'Languages of the Brain, The Neurophysiology of Remember-ing' *Scientific American*, January, NY
von Reibnitz,U. (1988) *Scenario Techniques*, McGraw-Hill, Hamburg
Rittel, H. and Webber, M. (1974) 'Dilemmas in General Theory of Planning' *Systems and Management Annual* 1974, Petrocelli, NY
Rucker, R. (1982) *Software*, Penguin Books, London
Rumelhart, D. (1986) *Parallel Distributed Processing*, A Bradford Book, MIT Press, London
Salk, J.E. (1983) *Anatomy of Reality*, Greenwood Publishing Group Inc., Westport, CT
Samuelson, K.(1977) *General Information Systems Theory in Design, Modelling and Development* Institutional paper, Informatics and Systems Science, Stockholm University
Samuelson, K., Borko, H. and Amey, G. (1977) *Information Systems and Networks*, North-Holland, Amsterdam
Schleicher, A. (1888) *Die Deutsche Sprache*, Cotta, Stuttgart
Schneider, S. and Boston, P (1993) *Scientists on Gaya*, MIT Press, London
Schumacher, E. (1973) *Small is Beautiful*, Harper & Row, NY
Schumacher, E. (1978) *A Guide for the Perplexed* Abacus, London
Shannon, C. and Weaver, W. (1964) *The Mathematical Theory of Communication*, University of Illinois Press, Urbana, USA
Simon, H. (1969) *The Sciences of the Artificial*, MIT Press, London

Simon, H. (1976) *Administrative Behavior*, The Free Press, NY
Skyttner, L. and Fagrell, F. (1988) 'Minimizing Human, Economic and Ecological Losses at Sea' TRITA IS-5152, Stockholm University
Skyttner, L. (1993) 'The Distress Signal as a System Function', *Kybernetes*, Vol. 22, 3, London
Smuts, J. (1973) *Holism and Evolution* (reprint), Greenwood Press, Westport CT
Stonier, T. (1990) *Information and the Internal Structure of the Universe*, Springer-Verlag, London
Taylor, A. (1973) in E. Laszlo's *The World System*, Braziller, NY
Thompson, F. (1951) in Geoffrey Cumberlege's *Poems of Francis Thompson*, Oxford University Press, Oxford
Toffler, A. (1980) *The Third Wave*, Collins, London
Tustin, A. (1955) *Automatic Control*, Simon & Schuster, NY
Ulrich, W. (1983) *Critical Heuristics of Social Planning*, Haupt, Berne
Warman, A. (1993) *Computer Security within Organizations* Macmillan Press, London
Watt, K. and Craig, P. (1988) 'Surprise, Ecological Stability Theory' in C. S. Holling's *The Anatomy of Surprise*, John Wiley, NY
Weawer, W.(1948) 'Science and Complexity', *American Scientist*, 36, 194, New Haven, Conn.
Weinberg, G. (1975) *An Introduction to General Systems Thinking*, John Wiley, NY
Wiener, N. (1948) *Cybernetics or Control and Communication in the Animal and the Machine*, John Wiley, NY
Wurman, R.S. (1991) *Information Anxiety*, Pan Books, London
Yovits, H. and Ernst, R. (1967) *Electronic Handbook of Information*, Thompson Washington, DC
Zuchov, G. (1979) *The dancing Wu-Li Masters*, Bantam Books, Toronto

Name Index

Subject Index